T0074728

Phytochemistry of Plants of Genus *Ocimum*

Phytochemical Investigations of Medicinal Plants

Series Editor:
Brijesh Kumar

Phytochemistry of Plants of Genus *Phyllanthus*
Brijesh Kumar, Sunil Kumar and K. P. Madhusudanan

Phytochemistry of Plants of Genus *Ocimum*
Brijesh Kumar, Vikas Bajpai, Surabhi Tiwari and Renu Pandey

Phytochemistry of Plants of Genus *Piper*
Brijesh Kumar, Surabhi Tiwari, Vikas Bajpai and Bikarma Singh

Phytochemistry of *Tinospora cordifolia*
Brijesh Kumar, Vikas Bajpai and Nikhil Kumar

Phytochemistry of Plants of Genus *Rauvolfia*
Brijesh Kumar, Sunil Kumar, Vikas Bajpai and K. P. Madhusudanan

Phytochemistry of *Piper betle* Landraces
Vikas Bajpai, Nikhil Kumar and Brijesh Kumar

For more information about this series, please visit: https://www.crcpress.com/Phytochemical-Investigations-of-Medicinal-Plants/book-series/PHYTO

Phytochemistry of Plants of Genus *Ocimum*

Brijesh Kumar, Vikas Bajpai, Surabhi Tiwari
and Renu Pandey

CRC Press is an imprint of the
Taylor & Francis Group, an **informa** business

First edition published 2020
by CRC Press
6000 Broken Sound Parkway NW, Suite 300, Boca Raton, FL 33487-2742

and by CRC Press
2 Park Square, Milton Park, Abingdon, Oxon, OX14 4RN

ISBN: 978-0-367-85753-0 (hbk)
ISBN: 978-1-003-01485-0 (ebk)

Typeset in Times
by codeMantra

Contents

Preface

Plants are considered as one of the most important sources of drugs used to alleviate and treat different ailments since the beginning of human civilization. Medicinal plants and their bioactive secondary metabolites have been considered as a fundamental source of medicine for the treatment of a range of diseases in our modern medical system. India has a rich heritage of medicinal plants, used in Ayurveda, Siddha, Unani, Homoeopathy and Naturopathy since the ancient era. In India, about 6000–7000 plant species are utilized in traditional, folk and herbal medicines. The *Ocimum* is one of the most recognized and used genera that includes more than 150 species of herbs and shrubs with immense medicinal properties. It is distributed throughout the tropical regions of Asia, Africa, and Central and South America. *Ocimum sanctum* (*Tulasi or Holy Basil*) is found throughout tropical and semitropical regions of India and other Asian countries. Different parts of this plant are traditionally utilized in the Indian system of medicine for the treatment of ailments.

Herbal medicines/formulations involve the use of crude or processed plant parts containing one or several active constituents. Due to the rapid commercialization of the herbal medicines, accurate knowledge of their phytochemical composition, which is directly related to medicinal properties, is a very crucial aspect for the safety and efficacy of the products. According to the European Medicines Evaluation Agency (EMEA) and Food and Drug Administration (FDA) regulations, identification and determination of the active constituents are prerequisites for the development of modern evidence-based phytomedicines. Nevertheless, the use of medicinal herbs or herbal drugs is increasing throughout the world. One of the major issues of its acceptance is the lack of quality control or standardization. Standardization is an essential step for the establishment of consistent pharmacological activity, a reliable chemical profile or simply a quality control program for the production of herbal drugs. Standardization of medicinal herbs or herbal products is necessary for assessment of their safety, efficacy and quality. Recently, mass spectrometry has earned a central role in the field of plant metabolomics due to its sensitivity, selectivity and accuracy. It facilitates efficient analysis of metabolites in the complex matrix of plant extracts due to high sensitivity, selectivity and versatility. Liquid chromatography (LC) coupled with tandem mass spectrometry (MS/MS) detection has increased separation of compounds, which makes this technique more sensitive, selective and suitable for qualitative and quantitative analyses of plant metabolites.

The work contained in this book includes reports of phytochemical investigations and study of biological potential of *Ocimum* spp. Metabolic profiling of six *Ocimum* species, namely, *O. americanum*, *O. basilicum*, *O. gratissimum*, *O.* kilimandscharicum, *O. sanctum* green and *O. sanctum* purple using HPLC-QTOF-MS/MS was successfully completed. Bioactive phenolic acids, flavonoids, propenyl phenol and terpenoids were simultaneously quantified in leaf extracts of *O. sanctum* from different geographical regions and its marketed herbal formulations using UHPLC-QqQ_{LIT}-MS/MS. Similarly, fifteen bioactive phytoconstituents were determined in leaf extracts of six *Ocimum* species. We hope this book may be useful to academicians, researchers, manufacturers and users of herbal products according to their aims and objective. It is expected that the effort will open new vistas of knowledge for multidisciplinary research in the coming years.

31/12/2019 **The Authors**

Acknowledgments

The completion of this book is due to the Almighty who blessed us with all the resources required to accomplish this journey. We are glad to have the opportunity to express our gratitude to Dr. K. P. Madhusudanan, whose constant encouragement, guidance and support helped us to complete the task successfully. We express our deep sense of gratitude to Director of CSIR-Central Drug Research Institute (CDRI), Lucknow, India, for his support and Sophisticated Analytical Instrument Facility (SAIF) Division of CSIR-CDRI, where all the data were generated.

Authors

Dr. Brijesh Kumar is a Professor (AcSIR) and Chief Scientist of Sophisticated Analytical Instrument Facility (SAIF) Division, CSIR-Central Drug Research Institute (CDRI), Lucknow, India. Currently, he is facility in charge at SAIF Division of CSIR-CDRI. He has completed his PhD from CSIR-CDRI, Lucknow (Dr. R.M.L Avadh University, Faizabad, UP, India). He has to his credit 7 book chapters, 1 book and 145 papers in reputed international journals. His current area of research includes applications of mass spectrometry (DART MS/Q-TOF LC-MS/ 4000 QTrap LC-MS/ Orbitrap MS^n) for qualitative and quantitative analyses of molecules for quality control and authentication/standardization of Indian medicinal plants/parts and their herbal formulations. He is also involved in the identification of marker compounds using statistical software to check adulteration/ substitution.

Dr. Vikas Bajpai completed his PhD from the Academy of Scientific and Innovative Research (AcSIR), New Delhi, India, and carried out his research work under the supervision of Dr. Brijesh Kumar at CSIR-CDRI, Lucknow, India. His research interest includes the development and validation of LC-MS/MS methods for qualitative and quantitative analyses of phytochemicals in Indian medicinal plants.

Ms. Surabhi Tiwari completed her masters in chemistry from the University of Allahabad, Prayagraj, India. She has worked in Pharmacognosy Division of National Botanical Research Institute, Lucknow, India, for the analysis of herbals using instruments like HPTLC and HPLC. Currently, she is working as a Senior Research Fellow in SAIF Division under the supervision of Dr. Brijesh Kumar at CSIR-Central Drug research Institute (CDRI), Lucknow, India. Her current research interest includes phytochemical analysis of medicinal plants.

Dr. Renu Pandey completed her PhD in Chemical Sciences from the Academy of Scientific and Innovative Research (AcSIR), New Delhi, India, and carried out her research work under the supervision of Dr. Brijesh Kumar at CSIR-Central Drug Research Institute (CDRI), Lucknow, India. Her research includes development and validation of LC-MS/MS methods for the qualitative and quantitative analyses of medicinal plants and derived herbal formulations/dietary supplements.

List of Abbreviations and Units

°C degree celsius
μL microliter
APCI atmospheric pressure chemical ionization
API atmospheric pressure ionization
BPC base peak chromatogram
CAD collision activated dissociation
CE capillary electrophoresis
CE collision energy
CID collision induced dissociation
CXP cell exit potential
Da dalton
DAD diode array detection
DP declustering potential
EP entrance potential
ESI electrospray ionization
FDA food and drug administration
FIA flow injection analysis
g gram
GC-MS gas chromatography-mass spectrometry
GS1 nebulizer gas
GS2 heater gas
h hour
HPLC high performance liquid chromatography
ICH international conference on harmonization
IS internal standard
IT ion trap
kPa kilopascal
L liter
LC liquid chromatography
LOD limit of detection
LOQ limit of quantification

LTQ	linear trap quadrupole
m/z	mass to charge ratio
mg	milligram
min	minute
mL	milliliter
mM	millimolar
MRM	multiple reaction monitoring
MS	mass spectrometry
ms	millisecond
MS/MS	tandem mass spectrometry
ng	nanogram
NMR	nuclear magnetic resonance
PCA	principal component analysis
PDA	photodiode array
psi	pressure per square inch
QqQ_{LIT}	hybrid linear ion trap triple quadrupole
QTOF	quadrupole time of flight
r^2	correlation coefficient
RDA	retro-Diels–Alder
RSD	relative standard deviation
S/N	signal to noise ratio
SD	standard deviation
t_R	retention time
UHPLC	ultra high performance liquid chromatography
UV	ultraviolet
WHO	world health organization
XIC/EIC	extracted ion chromatogram

Introduction 1

1.1 WORLDWIDE DISTRIBUTION OF GENUS *OCIMUM*

The Indian Himalaya is home to more than 8,000 species of vascular plants (Singh and Hajra 1996), of which 1748 are known for their medicinal properties (Samant et al. 1998). Higher plants have played key roles in the lives of tribal peoples living in the Himalaya by providing forest products for both food and medicine. Although numerous wild and cultivated plants have been utilized as curative agents since ancient times, they have gained more importance in recent days not only as herbal medicines but also as natural ingredients for the cosmetic industry. Plants have been used by humans from prehistoric times to get rid of suffering and cure various ailments. The folk medicines of almost around the world rely chiefly on herbal medicine even today. The therapeutic uses of plants are safe, economical, effective and easily available (Atal and Kapur 1982; Trivedi 2007; Joshi 2017). Ayurveda, the conventional system of medicine in India, is an indigenous, well known and explored system of alternative and complementary medicine. Dhanvantari, known as the physician of the God, is recognized as the founder of Ayurvedic medicine. Vedas, specially the Atharvaveda, mainly focus on curing the diseases through plant based remedies (Patwardhan et al. 2005). Species of genus *Ocimum*, family Lamiaceae, are prominently known as *Basil* (Chen et al. 2013). These are local to the tropical and warm mild areas and widely found in Asia, Africa, and Central and South America (Pandey et al. 2014; Beatovic et al. 2015). The genus *Ocimum* contains more than 150 species distributed throughout the tropical regions of Asia, Africa, and Central and South America (Darrah 1974; Vani et al. 2009; Prasad et al. 2012). *Ocimum* species are a rich wellspring of essential oils, the volatile liquid of aroma compounds extracted from the leaves and flowering tops of *basil* (Beatovic et al. 2015; Pandey et al. 2014). Essential oils extracted from different *Ocimum* species are used as flavoring or fragrance agents in foods and beverages, as well as in the pharmaceutical, perfumery and cosmetic industries (Beatovic et al. 2015; Pandey et al. 2014; Vieira et al. 2014). Even today, the home made remedy mostly depends on the homegrown medicinal plants. These plants are rich sources of medicines and expensive medicated supplements. The most well known components of home based medications are medicinal plants with lesser or no side effects. In the Indian system of medicine, Ayurveda, Homeopathy, Siddha and Unani together have taken the responsibility to deal with various ailments for centuries. *Ocimum sanctum* is known as the "Queen of Herbs" due to its tremendous medicinal properties that have been utilized in Ayurvedic system of medicine from the ancient period. Charaka, in his famous book *Charaka Samhita*, also mentioned about *Tulsi* and its various medicinal properties and actions. Five properties of *Tulsi*, namely, Rasa, Guna, Virya, Vipaka and Karma, are described in Ayurveda (Ayurvedic Pharmacopoeia of India; Anonymous 2016). It is a plant that is utilized for the religious as well as therapeutic purposes. The six species of *Ocimum* genus commonly found in India are *O. americanum* L. (syn. *O. canum* Sims), *O. basilicum* L., *O. gratissimum* L., *O. killimandscharicum* Baker ex Gürke, *O. sanctum* L. (syn. *O. tenuiflorum* L.) green and *O. sanctum* L. purple (Mahajan et al. 2013). Among these, *O. sanctum* (*holy basil*) is widely grown in India as a folk medicine and sacred plant, whereas *O. basilicum* (*sweet basil*) as a culinary and ornamental herb. Traditionally, *Tulsi* is also known as "the elixir of life" since it promotes longevity (Siddiqui 1993). The word "*Tulsi*" is from Sanskrit language which means "matchless" or the "incomparable one." In Hindu mythology, *Tulsi* is also known as Vaishnavi (belonging to Lord Vishnu), Vishnu Vallabha, Haripriya (beloved to Lord Vishnu) and Vishnutulsi. The famous holy book *Padma Purana* describes the importance of *Tulsi* with Lord Vishnu by regarding it as the incarnation (avatar) of Lord Lakshmi (wife of Lord Vishnu). Its leaves are important in the Hindu religion. The leaves are used in every holy rituals and ceremonies, and given as religious offering known as Prasad. It is also mentioned that lighting a single stick of *Tulsi* lamp is equal to enlighting lakhs of lamps for Lord Vishnu (Siddiqui 1993; Sen 1993). Most of the Hindus grow *Tulsi* in their household or at the entrance of their house as they believe it provides positive energy. They also believe that god dwells in *Tulsi* plant. Different parts of the plant are used in Ayurveda and Siddha systems of medicine for prevention and cure of many illnesses and everyday ailments like common cold, headache, cough, influenza, earache, fever, colic pain, sore throat, bronchitis, asthma, hepatic diseases, malarial fever, flatulence, migraine headaches, fatigue, skin diseases, wound, insomnia, arthritis, digestive disorders, night blindness and diarrhea, and they are also used as antidotes for snakebite and scorpion sting (Gupta et al. 2002; Ali et al. 2012; Mandal et al. 2012; Parag et al. 2010; Farivar et al. 2006; Shokeen et al. 2005; Sawarkar et al. 2010; Yucharoen et al. 2011; Suzuki et al. 2009; Saini et al. 2009; Deo et al. 2011;

Chandra et al. 2011). The tea made of *Tulsi* leaves removes toxins and infections from respiratory region and thus is good for headaches, coughs and kidney-related diseases (Bilal et al. 2012). *Tulsi* has a strong and powerful taste which helps in relieving cold and digestion problems, which are reported in Ayurveda as kapha and pitta doshas, respectively. The leaves are good for nerves and to sharpen memory (Batta and Santhakumari 1970). Chewing of *Tulsi* leaves also cures ulcers and infections of mouth (Prajapati et al. 2003; Verma 2016; Bhargava and Singh 1981). It also improves the tone of voice, provides strength, and gives a peaceful mind (Mahajan et al. 2013; Mohan et al. 2011; Pattanayak et al. 2010). Due to the presence of antioxidants and vitamins, especially vitamin A, in *Tulsi*, it is useful in cell repairing and strengthening (Gupta et al. 2002; Batta and Santhakumari 1970; Bhargava and Singh 1981). In India, *Tulsi* leaves mixed with water are used to liberate the soul from the body during death (Sen 1993; Khanna et al. 2003). *Tulsi* plantation in garden "tulsi-van" is the symbol of prosperity (Khanna et al. 2003). The dried leaves of *Tulsi* are used to protect grains from bacteria while storage. *Tulsi* acts as a purifier; for example, a few leaves dropped in drinking water or foodstuff can purify them, and thus they can easily be protected from germs and bacteria. During solar and lunar eclipses, *Tulsi* leaves are used in cooked and stored foods to prevent them from bacteria (Siddiqui 1993). *Tulsi* leaves are very much effective in boosting up the immune system. They protect from nearly all sorts of infections from viruses, bacteria, fungi and protozoa (Shukla et al. 2012). They are also used for controlling air pollution in most of the cities especially in Agra near the Taj Mahal (Mishra 2008). Modern studies had proven that they are also helpful in preventing or restricting the growth of tumor cells (Kumar et al. 2012; Verma 2016).

In India, plants have been traditionally used for human and veterinary health care, and also in the food and textile industries. Even though most of the resources for food, trade, cosmetics and perfumes of the local indigenous people were undocumented in literature, India occupies a special position in the field of herbal medicines. All parts of *O. sanctum* plants such as leaves, flowers, stem, root and seeds act as expectorant and analgesic; possess anticancer, anti-HIV, and antidiabetic activities; and act as hepatoprotective, antiulcer, antimicrobial, gastroprotective, antioxidant, hypotensive, hypolipidemic and antistress agents (Gupta et al. 2007; Pattanayak et al. 2010; Devi et al. 1998; Kelm et al. 2000; Asha et al. 2001; Hakkim et al. 2007; Mahajan et al. 2013; Mohan et al. 2011; Mondal et al. 2009; Beatovic et al. 2015; Prasad et al. 2012; Vieira et al. 2014; Runyoro et al. 2010; Gupta et al. 2013; Joshi 2013; Ramesh and Satakopan 2010; Eshraghian 2013; Ovesná et al. 2006). They also play a significant role in treatment of fevers, arthritis, convulsions, bronchitis, etc. in traditional medical practices (Nagarjun et al. 1989; Pandey et al. 2014; Bhattacharya 2004; Joshi 2017). Several commercial herbal preparations of

Ocimum are available in market which are used to treat various ailments such as cold, cough, nasal congestion and fever. Many herbal formulations of *Ocimum* reported in Ayurveda such as *Manasamitra Vatakam*, *Mahajvarankusa Rasa*, *Mukta Panchamrit Rasa and Tribhuvanakirti Rasa* are also easily available in market and can be easily prepared in home as household remedies (Anonymous 2016). The ongoing modern systematic research on *Tulsi* also shows that the oral consumption of *Tulsi* helps in curing psychological and physiological disorders. Plants of *Ocimum* species flourish well in different soils and various climatic conditions. The suitable soil for cultivation of *Ocimum* ranges from rich loam soil to pitiable laterite soil and soil having saline or alkaline to moderately acidic strength. Well drained soil boosts the growth of *Ocimum* plants and improves their vegetative growth. Basil thrives well under fair to high rainfall and humid environments and long sunny days with good sunlight. High temperature is favorable for the growth of *Ocimum* plants and increased secretion of essential oils from them.

1.2 INTRODUCTION: SELECTED *OCIMUM* SPECIES

In Indian culture, *O. sanctum* is worshipped as a goddess in homes and temples. *Ocimum* has more than 150 reported species. *O. sanctum*, *O. gratissimum (Ram tulasi*), *O. canum* (*Dulal tulasi*), *O. basilicum* (*Ban tulasi*), *O. kilimandscharicum*, *O. americanum*, *O. camphora* and *O. miranthum* are the most widely recognized and vital cosmopolitan plants. *O. sanctum* L. (*Tulsi*) is normally known as *holy basil* in English; *sacred basil* in Hindi; Vishnu Priya and Divya in Sanskrit; Tulasi in Assamese, Bengali, Kannada, Telugu and Malayalam; Tulsi and Tulasi in Gujarati and Punjabi; Raihan in Urdu; Thulasi and Tulasi in Tamil; and Tulas in Marathi (Anonymous 2016).

1.3 BOTANICAL DESCRIPTION OF SELECTED *OCIMUM* SPECIES

1.3.1 *O. americanum* L.

Habit: Aromatic annual or short lived perennial herb, 10–50(–80) cm tall; Stem: Rounded quadrangular or quadrangular, erect or ascending, sometimes with erect flowering shoots coming from a decumbent stem, woody at base, sometimes with epidermis peeling in strips, branched above, indumentum of short, appressed or retrorse hairs, becoming denser on the inflorescence axis, with or without sessile glands; Leaves: Sometimes folded along midrib on drying, with or without axillary fascicles of younger leaves, blade narrowly ovate or elliptic, 0.5–2.5 × 0.5–1.5 cm, entire to shallowly serrate, apex acute, base cuneate, glabrous to slightly pubescent above, pubescent with longer hairs on midrib and veins beneath, glandular-punctate; petiole 2–15 mm long, with appressed hairs and usually with longer patent hairs; Inflorescences: Lax, verticils ± 10 mm apart; axis densely pubescent with retrorse hairs; bracts often erect above forming a small coma around apex, deciduous or not, ovate, 3–4 mm long, entire, pilose; pedicels 1–2 mm long, ± erect, slightly flattened, curved; Calyx: 1.5–2 mm long at anthesis, indumentum of long patent hairs and sessile glands, interior of tube with a dense ring of hairs at throat, posterior lip rounded at tip, decurrent, median teeth of anterior lip lanceolate, acuminate, teeth of lateral lobes deltate, cuspidate, fruiting calyx 2–4 mm long, throat open, posterior lip accrescent, wider at tip, decurrent, lateral and median teeth of anterior lip ± convergent or not; Corolla: White or pale mauve, 4–5 mm long, tube straight, funnel-shaped; Stamens: Stamens exceeding corolla by 1–2 mm, posterior pair with fleshy, flattened, glabrous outgrowth near base; Ovary: Ovary glabrous; Fruits: Nutlets black, ovoid, longer than broad, 0.8–1 mm long, minutely tuberculate, mucilaginous when wet (Clayton, Flora of Tropical East Africa 1970).

1.3.2 *O. basilicum* L.

Habit: Aromatic, annual or short lived perennial herb, 20–60 cm tall; Stem: Stems rounded quadrangular, erect or ascending, often woody at base, branching above, glabrous or puberulent with minute hairs concentrated on two opposing faces of the stem, becoming minutely pubescent on the inflorescence axis, usually with young shoots in the axils of the leaves; Leaves: Leaf blades narrowly ovate to elliptic, 1.5–5 × 0.5–2 cm, entire to shallowly serrate or occasionally laciniate, apex acute to acuminate, base cuneate to attenuate, glabrous or with small hairs on veins beneath, glandular-punctate, petiole 2–40 mm long; Inflorescences: Inflorescence lax, verticils up to 12 mm apart, bracts deciduous or not, narrowly ovate to elliptical, 3–8 × 1–3 mm, acute to cuspidate at apex, cuneate at base; pedicel up to 3–4 mm long, erect, ± flattened, slightly curved; Calyx: Calyx ± downward pointing, 4–5 mm long at anthesis, posterior lip ± glabrous, tube and anterior lip pubescent or pilose, sparsely gland-dotted, interior with a dense ring of hairs at throat, posterior lip large, rounded at tip, decurrent, median teeth of anterior lip lanceolate, acuminate, lateral lobes deltate, cuspidate, fruiting calyx 6–8 mm long, throat open, posterior lip accrescent, decurrent, rounded and wider at tip, lateral and lower teeth of anterior lip ± convergent; Corolla: Corolla pink, white or creamy yellow, 7–8 mm long, tube straight, funnel-shaped, scarcely exceeding calyx; Stamens: Stamens exceeding corolla by 2–3 mm, posterior with a large, fleshy, flattened, glabrous outgrowth near base; Ovary: Ovary glabrous; Fruits: Nutlets black, ovoid, longer than broad, 2–2.5 mm long, ± smooth to minutely tuberculate, mucilaginous when wet (Clayton, Flora of Tropical East Africa 1970).

1.3.3 *O. gratissimum* L.

Habit: Aromatic perennial herb 0.6–2.5 m tall; Stem: Stems erect, rounded-quadrangular, much branched, often striate, woody at base with epidermis often peeling in strips, glabrous or with scattered patent hairs below, becoming pubescent at nodes and on the inflorescence-axis, or pubescent with ± retrorse and patent hairs becoming hoary or tomentose on the inflorescence axis; Leaves: Leaf blades often darker above, elliptic or ovate to broadly ovate, 1.5–15 × 1–8.5 cm, serrate, apex obtuse, acute or acuminate, base cuneate or attenuate, indumentum of scattered short hairs mostly confined to veins beneath, or lower surface pubescent with upper glabrous or pubescent overall, sometimes tomentose beneath, petiole 5–30 mm long, glandular-punctate or not; Inflorescences: Inflorescence lax or dense (calyces of adjacent verticils touching or verticils ± 1 cm apart), bracts erect above forming a small green terminal coma, becoming downward-pointing and deciduous towards base, ovate or narrowly ovate, 3–12 × 1–7 mm, acuminate, sometimes cucullate; pedicels 3–4 mm long, spreading or ascending, slightly curved, ± flattened; Calyx: Calyx horizontal, ± downward-pointing or strongly reflexed against the inflorescence axis, 2–3 mm long at anthesis, pubescent, hoary or tomentose, with sessile glands or not, posterior lip large, rounded at tip, decurrent, teeth of anterior lip lanceolate, acuminate, teeth of lateral lobes bristle pointed, fruiting calyx 6–8 mm long, tube closed, posterior lip accrescent, rounded or obtuse and wider at tip, decurrent, median lobes of anterior lip pressed against the posterior lip, teeth of the lateral lobes level with, or lower than, the underside of the anterior lip, sometimes spreading, indumentum pubescent, hoary or tomentose with sessile glands or not; Corolla: Corolla greeny or dull white or pale yellow, 3–4 mm long, tube straight, funnel-shaped, scarcely exceeding the calyx tube; Stamens: Stamens exserted 1 mm from corolla, posterior with a flattened, hairy or subglabrous appendage near base; Ovary: Ovary glabrous; Fruits: Nutlets brown, ± spherical, 1.5 mm in diameter, minutely tuberculate, producing a small amount of mucilage when wet (Clayton, Flora of Tropical East Africa 1970).

1.3.4 *O. kilimandscharicum* Gürke

Habit: Aromatic, perennial shrub to 2 m tall; Stem: Stems rounded-quadrangular, much branched, woody with epidermis sometimes peeling off in strips below, arising from a large woody rootstock, with long ± patent or antrorse, white eglandular hairs, becoming denser on the inflorescence axis, with sparse sessile glands; Leaves: Leaves often with fascicles of young leaves or young shoots in axils, blade ovate, (1–)1.5–5.5 × (0.5–)1–3 cm, serrate, apex obtuse or rounded, base cuneate; indumentum of long white appressed hairs or sometimes glabrous above, glandular-punctate; petiole 4–10 mm long; Inflorescences: Inflorescence lax, verticils 8–20 mm apart, bracts usually deciduous, forming a small coma or not, ovate, ± 3 mm long, 1–2 mm broad, entire, cuspidate, pedicel 3 mm long, erect, slightly curved, ± flattened; Calyx: Calyx ± downward-pointing, 3–4 mm long at anthesis, pilose to villose with the posterior lip less densely hairy, interior with a villose ring of hairs at throat, posterior lip rounded and wider at tip, decurrent, median teeth of anterior lip lanceolate, acuminate, teeth of lateral lobes deltate, cuspidate, fruiting calyx ± 5 mm long, throat open, posterior lip accrescent, wider at tip, decurrent; Corolla: Corolla white, pink or mauve, 6–7 mm long, tube straight, funnel-shaped, scarcely exceeding calyx tube; Stamens: Stamens exserted 3–5 mm, posterior with a fleshy, flattened, hairy outgrowth near base; Ovary: Ovary glabrous; Fruits:Nutlets black, longer than broad, 1.5 mm long, ± smooth, mucilagenous when wet (Clayton, Flora of Tropical East Africa 1970).

1.3.5 *O. tenuiflorum* L.

Habit: Aromatic herb or woody herb of 1 m tall; Stem: Stems erect, branched, woody at base, indumentum of patent hairs; Leaves: Leaves petiolate, blade broadly elliptical, 1.5–3.3 × 1.1–2 cm, serrate, apex obtuse, base cuneate, with an indumentum of short, appressed hairs, petiole 7–15 mm long; Inflorescences: Inflorescence lax, pedicels in fruit 3–4 mm long, spreading; Calyx: Calyx 1(–3 in fruit) mm long, throat open and glabrous; Corolla: Corolla pink or white, 3 mm long, tube ± parallel sided; Stamens: Posterior stamens ciliate, cilia concentrated on one side of the filament near base; Fruits: Nutlets brown, ovoid, ± 1 mm long, producing small amounts of mucilage when wet (Clayton, Flora of Tropical East Africa 1970).

1.4 PHYTOCHEMISTRY OF SELECTED *OCIMUM* SPECIES

More than 200 phytochemicals have been reported from *Ocimum* species since 1930, which shows a vast diversity of phytochemicals among the *Ocimum* genus. Principal phytochemicals reported are phenolic compounds including phenolic acids, propenyl phenols, flavonoids and terpenoids (Kelm et al. 2000; Grayer et al. 2001; Grayer et al. 2002; Runyoro et al. 2010; Joshi 2013; Mahajan et al. 2013). Table 1.1 lists some important phytochemicals reported from these species.

TABLE 1.1 List of Phytochemicals Reported from Different *Ocimum* Species

S. NO.	*PLANT SPECIES*	*PHYTOCHEMICALS*	*REFERENCES*
1.	*O. basilicum* L.	Methyl chavicol, linalool, eugenol	Mondawi et al. 1984
2.	*O. gratissimum* L.	Thymol, *p*-cymene, eugenol, γ-terpinene	Ntezuriubanza et al. 1987
3.	*O. basilicum* L.	Linalool, eugenol, methyl eugenol, fenchyl alcohol	Akgül 1989
4.	*O. utricifolium* L.	Trans-methylisoeugenol, trans-β-ocimene, *cis*-β-ocimene, methyleugenol, eugenol, methyl chavicol, linalool	Janseen et al. 1998
5.	*O. basilicum* L.	Methyl chavicol, linalool, Methyl eugenol, β-caryophyllene, α-pinene, β-pinene, limonene, camphene	Khatri et al. 1995
6.	*O. basilicum* L.	Linalool, methyl chavicol, eugenol	Marotti et al. 1996
7.	*O. basilicum* L.	Limonene, linalool	Zollo et al. 1998
8.	*O. canum*	1,8-Cineole, β-pinene	Chalchat et al. 1999
9.	*O. gratissimum*	p-Cymene, γ-terpinene, thymol	Chalchat et al. 1999

(Continued)

TABLE 1.1 (*Continued*) List of Phytochemicals Reported from Different *Ocimum* Species

S. NO.	*PLANT SPECIES*	*PHYTOCHEMICALS*	*REFERENCES*
10.	*O. basilicum*	Linalool, methyl eugenol	Chalchat et al. 1999
11.	*O. basilicum*	Linalool, eugenol, (E)-α-bergamotene, thymol	Kéita et al. 2000
12.	*O. americanum* L.	Geranial, neral, camphor, limonene	Mondello et al. 2002
13.	*O. basilicum* var. *Purpurascens*	Linalool, geranial, geranyl acetate	Silva et al. 2008
14.	*O. sanctum* L.	Eugenol, limonene, E. caryophyllene, eugenol	Silva et al. 2008
15.	*O. basilicum*	1,8-Cineole, linalool, estragole	Silva et al. 2008
16.	*O. basilicum*	Linalool, α-muurolol methyl eugenol, α-cubebene, nerol, ε-muurolene	Özcan et al. 2002
17.	*O. minimum*	Geranyl acetate, terpinen-4-ol, octan-3-ylacetate	Özcan et al. 2002
18.	*O. gratissimum* L.	Eugenol, 1,8-cineole	Silva et al. 2008
19.	*O. micranthum Willd.*	Eugenol, (E)-β-caryophyllene, bicyclogermacrene	Tchoumbougnang et al. 2005
20.	*O. selloi Benth.*	Anethole, linalool	Tchoumbougnang et al. 2005
21.	*O. gratissimum*	β-Phellandrene, limonene, γ-terpinene, thymol	Tchoumbougnang et al. 2005
22.	*O. basilicum*	Linalool, methyl chavicol, methyl cinnamate	Kasali et al. 2005
23.	*O. basilicum*	Linalool, 1,8-cineole, eugenol	Ismail 2006
24.	*O. basilicum*	Methyl chavicol, linalool, epi-α-cadinol, trans-αbergamotene	Sajjadi 2006
25.	*O. basilicum*	Methyl chavicol, geranial, neral, caryophyllene oxide	Awasthi and Dixit 2007
26.	*O. sanctum*	Methyl eugenol, E-caryophyllene	Awasthi and Dixit 2007
27.	*O. sanctum*	Rama eugenol, E-caryophyllene, β-elemene	Bunrathep et al. 2007

(*Continued*)

TABLE 1.1 (*Continued*) List of Phytochemicals Reported from Different *Ocimum* Species

S. NO.	*PLANT SPECIES*	*PHYTOCHEMICALS*	*REFERENCES*
28.	*O. sanctum*	β-Ocimene, methyl chavicol	Bunrathep et al. 2007
29.	*O. canum*	Linalool, neral, geranial, β-caryophyllene	Bunrathep et al. 2007
30.	*O. gratissimum*	Z-β-ocimene, β-caryophyllene, γ-muurolene, α-farnasene, eugenol	Bunrathep et al. 2007
31.	*O. sanctum*	β-Caryophyllene, methyl eugenol	Bunrathep et al. 2007
32.	*O. basilicum*	1,8-Cineole, linalool, estragole, (E)-methyl cinnamate	Politeo et al. 2007
33.	*O. campechianum*	Eugenol, methyl eugenol, 1,8-cineole, elemicin	Zoghbi et al. 2007
34.	*O. gratissimum*	Thymol, γ-terpinene, 1,8-cineole, *p*-cymene	Zoghbi et al. 2007
35.	*O. basilicum*	Linalool, methyl chavicol	Zheljazkov et al. 2007
36.	*O. basilicum*	Estragole, linalool	Koba et al. 2009
37.	*O. basilicum*	Methyl chevicol, linalyl acetate, camphene	Anand et al. 2011
38.	*O.kilimandscharicum*	Camphor, DL-limonene	Anand et al. 2011
39.	*O. gratissimum*	Eugenol, *cis*-ocimene, germacren	Anand et al. 2011
40.	*O. gratissimum*	Eugenol, (Z)-β-ocimene, germacrene-D	Verma et al. 2011
41.	*O.kilimandscharicum*	Camphor, limonene, camphene, γ-terpinene	Verma et al. 2011
42.	*O. basilicum*	Methyl cinnamate, linalool, tau-cadinol	Mohammed et al. 2012
43.	*O. basilicum*	Methyl chavicol (62.00), linalool	Sastry et al. 2012
44.	*O. gratissimum*	Eugenol, limonene 1,8-cineole	Sastry et al. 2012
45.	*O. tenuiflorum*	Eugenol, β-caryophyllene	Sastry et al. 2012

1.5 PHARMACOLOGICAL ACTIVITY OF SELECTED *OCIMUM* SPECIES

Ocimum species have been reported to exhibit various pharmacological activities such as anticancer, antidiabetic, antiinflammatory, anti-HIV, antimicrobial, antioxidant, antistress, antiulcer, cardioprotective, gastroprotective, hepatoprotective and immunomodulatory (Devi et al. 1998; Kelm et al. 2000; Asha et al. 2001; Hakkim et al. 2007; Gupta et al. 2007; Pattanayak et al. 2010; Mahajan et al. 2013). Some of the reported biological activities are shown in Table 1.2.

TABLE 1.2 Pharmacological Activities of Selected *Ocimum* Species

S. NO	*ACTIVITY*	*OCIMUM SPECIES*	*REFERENCES*
1.	Anthelmintic activity	*O. gratissimum*	Pessoaa et al. 2002
2.	Antibacterial activity	*O. gratissimum*	Nakamura et al. 1999
3.	Antimicrobial activity	*O. gratissimum*	Okigbo et al. 2005
4.	Antidiabetic activity	*O. sanctum*	Parasuraman et al. 2015
5.	Antilipidemic activity	*O. basilicum*	Pandey and Anita 1990
6.	Eye disease	*O. sanctum*	Patil et al. 2011
7.	Mosquitocidal activity	*O. sanctum*	Kelm and Nair 1998
8.	Anticancer activity	*O. sanctum*	Nakamura et al. 2004
9.	Antibacterial activity	*O. sanctum*	Mishra and Mishra 2011
10.	Antifertility activity	*O. sanctum*	Batta and Santhakumari 1970
11.	Antimicrobial activity	*O. basilicum*	Ba-Hamdan et al. 2014
12.	Antioxidant activity	*O. gratissimum*	Akinmoladun et al. 2007
13.	Antiinflammatory, analgesic and antipyretic activities	*O. sanctum*	Kumar et al. 2015
14.	Cardioprotective activity	*O. basilicum*	Fathiazad et al. 2012
15.	Antioxidant activity	*O. basilicum*	Fathiazad et al. 2012
16.	Insecticidal activity	*O. gratissimum*	Adeniyi et al. 2010

The distinctive parts of *Ocimum* plants, such as leaves, roots and flowers, are utilized as customary remedies for nausea, flatulence, cold, dysentery, mental fatigue, spasm, rhinitis, arthritis, malaria, diarrhea, skin diseases, conjunctivities, bronchitis, bringing down glucose level, and recuperating wounds (Runyoro et al. 2010; Ramesh and Satakopan 2010; Mahajan et al. 2013; Gupta et al. 2013). Consumption of basil tea has been proved to cure coughs, migraines, diseases of upper respiratory tract and kidney, as well as to remove toxic materials from body (Bilal et al. 2012).

Several *Ocimum* species showed different biological activities performed on microorganisms (Table 1.3). Many *Ocimum* species are reported worldwide for their antibacterial, anticancer, antidyspepsia, antifungal, antigiardial, anti-inflammatory, antioxidant, antiproliferative, antiulcer, antiviral, antiwormal, insecticidal, wound healing and central nervous system (CNS) stimulant properties, and they also show hypoglycemic and hypolipidemic effects, cardiac stimulant response, and inhibitory effect on platelet aggregation. Studies by various groups for antibacterial and antifungal activities of *Ocimum* plants against a broad range of gram-negative and gram-positive bacteria, yeast and mold showed significant antimicrobial activities. Ethanolic, methanolic and hexane extracts of *O. sanctum* were analyzed for their *in vitro* antibacterial and antifungal activities; among these, the ethanolic extract showed the activity against only 9 strains, whereas methanolic and hexane extracts showed against 11 and 13 strains. The ethanolic extracts of *O. sanctum* seeds also showed noteworthy activities against ulceration in animal models. The ethanolic and aqueous extracts of *O. sanctum* showed significant activities against viruses such as coxsackie virus B1 and entero virus. Bhatti (2008) also reported antiwormal responses of ethanolic extracts of *O. sanctum*.

When On treatment of haemophilus influenzae with crude basil essential oils and by the phytochemicals isolated from essential oils of basil, approximately 81% of rats infected with *Haemophilus influenzae* and 75% of rats infected with *Streptococcus pneumoniae* (Pneumococci) were cured, respectively (Kristinsson et al. 2005). The formulations consisted of phytochemicals menthol and menthone obtained from vapors of peppermint oil and phytoconstituents linalool and eugenol from essential oil of *sweet basil* were tested against *Sclerotinia sclerotiorum* (Lib.) and *Rhizopus stolonifer* (Ehrenb. ex Fr.) by Edris et al. (2003). The phytoconstituent linalool from the essential oil of *sweet basil* showed a moderate antifungal activity, whereas the other phytochemicals like eugenol showed no activity at all. Fifteen compounds, representing 74.19% of the total essential oil content of the aerial parts of *O. basilicum*, obtained by hydrodistillation showing significant antifungal activity against some plant pathogenic fungi were identified by GC-MS (De Martino et al. 2009).

TABLE 1.3 *Biological Activities of Ocimum Extracts/Constituents Performed on Various Microorganisms*

S. NO.	*STUDIED ORGANISM FOR ACTIVITY*	*OCIMUM CONSTITUENTS/ EXTRACTS*	*REFERENCES*
Antibacterial activities			
1.	*Staphylococcus aureus*, *Salmonella enteritidis*, *Escherichia coli*	Essential oil	Anonymous 2010
2.	*Bacillus cereus*	Essential oil	Budka and Khan 2010
3.	*Pseudomonas aeruginosa*, *Listeria monocytogenes*, *Shigella* sp., *S. aureus*	Methanol extract	Kaya et al. 2008
4.	Two different strains of *E. coli*	Methanol extract	Kaya et al. 2008
5.	*H. influenzae* and *Pneumococci*	Essential oil	Kristinsson et al. 2005
6.	*S. aureus*	Essential oil	Nguefack et al. 2004
7.	*Staphylococcus*, *Enterococcus* and *Pseudomonas*	Linalool, methyl chavikol, methyl cinnamat and linolen	Opalchenova and Obreshkova 2003
8.	*Escherichia coli*, *P. aeruginosa*, *S. typhi*, and *S. aureus.*	Extract	Onwuliri et al. 2006
9.	*S. aureus*, *E. coli*, *B. subtilis*, *Pasteurella multocida* and pathogenic fungi *Aspergillus niger*, *Mucor mucedo*, *Fusarium solani*, *Micrococcus*, gram-negative and gram-positive	Essential oil	Suppakul et al. 2003a; Suppakul et al. 2003b
10.	*Giardia lamblia*	Linalool	de Almeida et al. 2007
11.	*P. aeruginosa*, *Acinetobacter*, *Bacillus*, *Escherichia*, *Staphylococcus*	Ethanol extract	Adigüzel et al. 2005
12.	*Acinetobacter*, *Bacillus*, *Brucella*, *Escherichia*, *Micrococcus* and *Staphylococcus*	Methanol extract	Adigüzel et al. 2005

(*Continued*)

TABLE 1.3 (*Continued*) *Biological Activities of Ocimum Extracts/Constituents Performed on Various Microorganisms*

S. NO.	*STUDIED ORGANISM FOR ACTIVITY*	*OCIMUM CONSTITUENTS/ EXTRACTS*	*REFERENCES*
13.	*Acinetobacter*, *Bacillus*, *Brucella*, *Escherichia*, *Micrococcus*, *Staphylococcus*	Hexane extract	Adigüzel et al. 2005
14.	*B. cereus*, *B. subtilis*, *B. megaterium*, *E. coli*, *S. aureus*, *L. monocytogenes*, *Shigella boydii*, *S. dysenteriae*, *Vibrio mimicus*, *V. parahaemolyticus* and *S. typhi*	Essential oils, methanol extract	Hossain et al. 2010a
15.	*S. dysenteriae*	Essential oils, methanol extract	Bassolé et al. 2010
Antifungal activity			
16.	*S. sclerotiorum* (Lib.), *R. stolonifer* (Ehrenb. exFr.) *Vuill*, *Mucor* sp. (Fisher)	Linalool and eugenol	Edris et al. 2003
17.	*Alternaria alternate* (Fries: Fries) *von Keissler Fulvia fulva* (Cooke) *Ciferri*, *F. solani* var. *coeruleum*, *Glomerella cingulate* (Stonem.) Spauld.	Essential oil	Zhang et al. 2009
18.	*F. oxysporum* f. sp. *vasinfectum* and *R. nigricans*	Cineol, linalool, methyl chavicol and eugeno	Reuveni et al. 1984
19.	Three isolates of *Candida albicans*	Methanol and hexane extract	Adigüzel et al. 2005
20.	*C. albicans*, *Penicillium notatum*, *Microsporeum gyseum*	Essential oil	Anonymous 2010
21.	Yeast and mold.	Essential oil	Suppakul et al. 2003a; Suppakul et al. 2003b
22.	*Aureobasidium pullulans*, *Debaryomyces hansenii*, *P. simplicissimum*, *P. citrinum*, *P. expansum*, *P. Aurantiogriseum*	Essential oil	De Martino et al. 2009
Antiseptic activity			
23.	*Proteus vulgaris*, *B. subtilis*, *S. paratyphi*	Essential oil	Anonymous 2010

(*Continued*)

TABLE 1.3 (*Continued*) *Biological Activities of Ocimum Extracts/Constituents Performed on Various Microorganisms*

S. NO.	*STUDIED ORGANISM FOR ACTIVITY*	*OCIMUM CONSTITUENTS/ EXTRACTS*	*REFERENCES*
Antiviral activity			
24.	Herpes virus, adeno viruses, hepatitis B virus and coxsackie virus B1 and entero virus 71	Apigenin, linalool	Pavela et al. 2004
Insecticidal activity			
25.	*Spodoptera littoralis* (Egyptian cottonworm)	Extract	Pavela et al. 2004
26.	*Tribolium castaneum*, *Sitophilus oryzae*, *Stegobiom paniceum*, *Bruchus chinensis* (stored grain insects)	Ocimene, cineole, linalool, methyl cinnamate, methyl chavicol	Bhatti 2008
27.	*Ceratitis capitata*, *Bactrocera dorsalis*, *B. cucurbitae*	Basil oil and trans-anethole, estragole, linalool	Ling Chang et al. 2009
Repellent activity			
28.	*Culex pipiens*	Essential oils	Pavela et al. 2004
29.	Mosquito	Essential oils	Erler et al. 2006
30.	Epimastigotes and trypomastigotes forms	Essential oil eugenol and linalool	Santoro et al. 2007

O. basilicum L. is a typical herb developed in plane zones as a decorative plant. The whole plant has its therapeutic importance; however, its leaves are particularly utilized for household remedies. The leaves are utilized to treat coughs, ear and bronchitis infections, and irritations, while the seeds are utilized to treat gonorrhea malady and weakness. Root extract is used to treat malarial fever. *Ocimum* belongs to Lamiaceae (syn. Labiatae) family, which is unique among the most imperative essential oil containing families, having in excess of 252 genera and 7,000 species in the plant kingdom. This family is known for its restorative properties and clinical trials and researches are being conducted to provide evidence for its restorative and remedial biological properties. *Ocimum* genus has a number of subspecies due to polymorphism, and these subspecies have variations in their chemical composition.

The oil of *O. basilicum* has shown the highest antiproliferative activity against the murine leukemia (P388) cell line, which makes it a potential plant for cancer treatment. It also helps in relieving functional dyspepsia in young female patients with dysmotility. De Almeida et al. (2007) have reported antigiardial activity of the *O. basilicum* essential oils and its purified substances. The methanolic extract of the *O. basilicum* showed antiinflammatory activity. Polyphenols isolated from the methanolic extract of *O. basilicum* were identified for antioxidant activity, and excellent synergistic effect was also observed against α-tocopherol.

Two phenolic compounds, rosmarinic and caffeic acids isolated from *sweet basil* also showed antioxidant activities. Essential oils of five species of the genus *Ocimum*, namely, *O. basilicum*, *O. basilicum* var. *purpurascens*, *O. gratissimum*, *O. micranthum* and *O. tenuiflorum*, were extracted by the steam hydrodistillation method. These oils were tested using the hypoxanthine/xanthine oxidase assay; although strong antioxidant activities were observed in all the oils, the highest antioxidant effect was found in the oil of *O. tenuiflorum* (IC50 = 0.46 μL/mL) when compared with that of *O. basilicum* var. *purpurascens* (IC50 = 1.84 μL/mL). The highest antioxidant activity of a methanolic extract of *O. basilicum* was also reported from different *in vitro* assay model systems such as the 2,2-diphenyl-1-picryl-hydrazyl-hydrate (DPPH) scavenging assay system and the oxidation of the soy phosphotidylcholine liposome model system. Phenolic compounds responsible for the antioxidative activity of the fractions were characterized and reported by liquid chromatography atmospheric pressure chemical ionization mass spectrometry. Ethanolic extracts of aerial parts of *O. basilicum* were reported to have cardiac effects. The alcoholic extracts exhibited a cardiotonic effect and the aqueous extract indicated a β adrenergic effect. The essential oil of *O. basilicum* was also reported for CNS activities, namely, sedative, hypnotic, anticonvulsant and local anesthetic. Zeggwagh et al. reported that the aqueous extract from whole plant of *O. basilicum* showed hypoglycemic effect in normal and streptozotocin diabetic rats. After a single oral administration, the extract of *O. basilicum* significantly reduced blood glucose levels in normal ($p < 0.01$) and diabetic rats ($p < 0.001$). After 15 days of repeated oral administration, *O. basilicum* produced a potent reduction in blood glucose levels ($p < 0.001$) in diabetic rats and a less reduction in normal rats ($p < 0.05$). In addition, plasma insulin levels and body weight remained unchanged over 15 days of oral administration in normal and diabetic rats. They also reported the hypocholesterolemic and hypotriglyceridemic activities of the aqueous extract of basil in hyperlipemic rats that were induced by high fat diet. It showed significant decrease in plasma and liver total cholesterol ($p < 0.02$ and $p < 0.05$, respectively) and triglyceride ($p < 0.02$ and $p < 0.01$, respectively). Similar result was observed on plasma low density lipoprotein cholesterol concentrations ($p < 0.02$). The aqueous extract of *O. basilicum* whole plant significantly reduced the cholesterol and triglycerides levels after repeated oral administration in diabetic rats ($p < 0.001$ and $p < 0.05$, respectively). *O. basilicum* has a potential antithrombotic profile *in vivo* as it showed an inhibitory effect on platelet aggregation induced by adenosine diphosphate (ADP) and thrombin.

O. kilimandscharicum, commonly known as *African blue basil*, is known for its camphor like fragrance. Similarly, *O. minimum* and *O. citriodorum* are used in cosmetic industry especially in Indonesia, Mexico and African countries. Essential oils extracted from *Ocimum* species are also used in food industry and aromatherapy. Antimicrobial, adaptogenic, antidiabetic, hepatoprotective, antiinflammatory, anticarcinogenic, radioprotective, immunomodulatory, neuroprotective, cardioprotective and mosquito repellent properties of *O. sanctum* have been reported in the literature.

Ocimum species are also used in traditional Iranian medicine as a culinary herb and a well known source of flavoring principles. Quantification of phenolic acids in these plants, which was determined using high performance liquid chromatography, showed drastic variations between different species. Chemical studies also revealed that rosmarinic acid is the predominant phenolic acid present in both flowers and leaves of plants from *Ocimum* species. There is very little literature available on the toxicity of *Ocimum* spp. However, *O. basilicum*, used as the most medicinally active herb, has been the most analyzed species. It was reported to contain several potentially dangerous compounds, namely, safrole, rutin, caffeic acid, tryptophan and quercetin. *p*-Coumaric acid and caffeic acid (phenolic acids) inhibit the digestion of plant cell walls in ruminants, because of their antimicrobial activity. When these phenolic acids are metabolized by rumen microbes, benzoic acid, 3-phenylpropionic acid and cinnamic acid are formed. When these compounds are detoxified, hippuric acid is formed. 3-Phenylpropionic acid can decrease the metabolic efficiency. Quercetin, a reported flavonoid from all *Ocimum* species, is a cocarcinogen in bracken fern (*Pteridium aquilinum*) which also interacts with Bovine papilloma virus type 4, leading to malignant epithelial papillomas in the upper alimentary tract. Safrole has been reported to cause cancer in rats. *Ocimum* oil also contains *d*-limonene, which has carcinogenic properties.

Ocimum species are characterized by variations in their morphology such as shape, size and pigmentation of leaves, which cause differences in their chemical composition and affect the commercial value of this genus (Ntezurubanza et al. 1987; Akgül 1989). Therefore, an efficient and reliable method is required for rapid screening and determination of phenolics and triterpenic acids in leaf extracts of *Ocimum* species and to study their interspecies variation.

FIGURE 1.1 Pictures of selected *Ocimum* species.

The literature survey indicated a variety of analytical methods for phytochemical analysis of *Ocimum* species such as high performance liquid chromatography (HPLC), high performance thin layer chromatography (HPTLC), gas chromatography-mass spectrometry (GC-MS), attenuated total reflectance/Fourier transform-infrared (ATR/FT-IR), FT-Raman and near-infrared (NIR) spectroscopy (Hakkim et al. 2007; Choudhury et al. 2011; Sundaram 2011; Thyagaraj et al. 2013; Anandjiwala et al. 2006; Lalla et al. 2007; Alam et al. 2012; Grayer et al. 2001; Sarkar et al. 2012; Sundaram et al. 2012; Khatri et al. 1995; Ntezurubanza et al. 1987; Zollo et al. 1998; Kéita et al. 2000; Schulz et al. 2003; Silva et al. 2008; Dey et al. 1983; Kitchlu et al. 2013; Kothari et al. 2004; Laskar et al. 1988; Vani et al. 2009; Rout et al. 2012; Srivastava et al. 2013; Itankar et al. 2015). These methods have various drawbacks such as low sensitivity, low resolution, high solvent consumption, long analysis time and need of derivatization.

Recently, liquid chromatography tandem mass spectrometric (LC-MS/MS) techniques have demonstrated to be more effective for rapid screening and determination of plant metabolites than the reported methods due to their high sensitivity, selectivity, specificity and shorter analysis time (Wu et al. 2013). So far, few LC-MS/MS methods using triple quadrupole and ion trap mass spectrometers were reported for phytochemical analysis of *Ocimum* species (Figure 1.1), but these are focused on the screening of flavonoids only (Grayer et al. 2000; Grayer et al. 2001).

Rapid Screening of Phytochemicals in *Ocimum* Species 2

2.1 PLANT MATERIAL

The plant materials (leaves of *Ocimum americanum*, *O. basilicum*, *O. gratissimum*, *O. kilimandscharicum*, *O. sanctum* green and *O. sanctum* purple) were collected from Nauni, Solan, Himachal Pradesh, India. Voucher specimens of *O. americanum*-8878 (**1**), *O. basilicum*-8879 (**2**), *O. gratissimum*-13422 (**3**), *O. kilimandscharicum*-8869 (**4**), *O. sanctum* green-11602 (**5**) and *O. sanctum* purple-8871 (**6**) have been deposited in the Department of Forest Products, Dr. Yashwant Singh (Y. S.) Parmar University of Horticulture and Forestry, Nauni, Solan, Himachal Pradesh, India.

2.2 INSTRUMENTATION AND ANALYTICAL CONDITIONS

2.2.1 HPLC-QTOF-MS/MS Conditions

The HPLC-QTOF-MS/MS analysis was performed on an Agilent 1200 HPLC system (Agilent Technologies, USA) connected to an Agilent 6520 QTOF-MS/MS system via a dual ESI interface. HPLC instrument composed of a quaternary pump (G1311A), online vacuum degasser (G1322A), autosampler (G1329A), thermostatted column compartment (G1316C) and diode array detector (G1315D). The HPLC separation was accomplished on a Thermo Betasil C_8 column (250 mm × 4.6 mm id, 5 μm) operated at 25°C. A gradient elution was achieved using two solvents: 0.1% (v/v) formic acid aqueous solution (A) and acetonitrile (B) at a flow rate of 0.4 mL/min. The 65 min HPLC gradient elution program was as follows: 25%–55% (B) from 0 to 15 min, 55%–55% (B) from 15 to 25 min, 55%–65% (B) from 25 to 30 min, 65%–75% (B) from 30 to 40 min, 75%–90% (B) from 40 to 59 min, 90%–25% (B) from 59 to 65 min and equilibration time 5 min. The sample injection volume was 4 μL.

Nitrogen was used as drying and collision gas in the ESI source. The ion source parameters were as follows: drying gas flow rate, 12 L/min; heated capillary temperature, 350°C; nebulizer pressure, 45 psi; VCap, fragmentor, skimmer and octapole RF peak voltages set at 3,500, 150, 65 and 75 V, respectively. The detection was carried out in positive and negative electrospray ionization modes, and spectra were recorded by MS scanning in the range of *m/z* 50–1,000. The MS/MS analyses were carried out by targeted fragmentation, and collision energy was set at 8–40 eV. Mass Hunter software version B.04.00 build 4.0.479.0 (Agilent Technologies) was used to control LC-MS/MS system, data acquisition and processing, including the prediction of chemical formula and exact mass calculation.

2.3 HPLC-QTOF-MS/MS ANALYSIS

Metabolic profiling of leaf extract of *Ocimum* species revealed the identification of fifty phytochemicals including twenty eight flavonoids, four propenyl phenol subsidiaries, two triterpenic acids, eleven phenolic acids and five phenolic acid esters. The structural identification of each compound was successfully completed based on their exact mass, molecular formula and MS/MS fragmentation by HPLC-ESI-QTOF-MS/MS. Twenty three phytochemicals were unambiguously identified and characterized by comparing their retention time and fragmentation pattern with authentic standards. Other compounds were tentatively identified by comparing their MS and MS/MS information with available literature. All the identified phytochemicals along with retention time, molecular formulas, *m/z* calculated and observed, error (Δ ppm), MS/MS data and their comparative profile for six *Ocimum* species are presented in Table 2.1. The HPLC-QTOF-MS/MS analysis of flavonoids, propenyl phenol derivatives and triterpenic acids was done in positive

TABLE 2.1 Compounds Identified from Leaf Extracts of *Ocimum* Species in Positive Ionization Mode by HPLC-QTOF-MS/MS

PEAK NO.	T_R *(MIN)*	*MOLECULAR FORMULA*	*CALC. M/Z [M+H]+*	*OBS. M/Z [M+H]+*	*ERROR (Δ PPM)*	*MS/MS DATA M/Z (% ABUNDANCE)*	*CE*	*IDENTIFICATION*	*DETECTION*	*CLASS*
1	7.5	$C_{21}H_{20}O_{11}$	449.1078	449.1079	0	431.0958 (16.1), 413.0843 (29), 395.0732 (24), 383.0760 (18.2), 365.0638 (12.3), 353.0628 (44.7), 339.0837 (16.3), 329.0644 (100), 313.0674 (5.2), 299.0538 (89.1), 287.0528 (8), 243.0334 (2), 217.0459 (1.3), 137.0254 (1.8)	25	Isoorientin*	OA, OG, OK, OS, OTF	Flavonoid C-glycoside
2	8	$C_{27}H_{30}O_{16}$	611.1607	611.1607	0.01	465.1013 (14.6), 449.1031(1.50), 345.0593 (0.4), 303.0485 (100), 287.0547 (0.3), 255.0830 (0.3), 147.0639 (4.4), 129.0533 (5.7), 85.0287 (4.2)	11	Quercetin-3-*O*-rutinoside (rutin)*	OA, OB, OG, OK, OS, OTF	Flavonoid C-glycoside
3	8.2	$C_{21}H_{20}O_{11}$	449.1078	449.1077	0.23	431.0975 (36.6), 413.0869 (50.4), 395.0764 (18.6), 383.0768 (23), 367.0817 (10.8), 353.0661 (23.9), 339.0832 (13), 329.0658 (100), 311.0550 (8.6), 299.0548(43.7), 287.0562 (2.4), 259.0622 (2.5), 217.0495 (3.1), 137.0212 (1.6)	25	Orientin*	OA, OG, OK, OS, OTF	Flavonoid C-glycoside

(*Continued*)

TABLE 2.1 (*Continued*) Compounds Identified from Leaf Extracts of *Ocimum* Species in Positive Ionization Mode by HPLC-QTOF-MS/MS

PEAK NO.	*T_R (MIN)*	*MOLECULAR FORMULA*	*CALC. M/Z $[M+H]^+$*	*OBS. M/Z $[M+H]^+$*	*ERROR (Δ PPM)*	*MS/MS DATA M/Z (% ABUNDANCE)*	*CE*	*IDENTIFICATION*	*DETECTION*	*CLASS*
4	9	$C_{21}H_{20}O_{10}$	433.1129	433.1127	0.5	415.1006 (12), 397.0895 (27.3), 379.0801 (32.7), 367.0791 (26.7), 361.0700 (13.7), 355.0772 (1), 351.0861 (7.3), 349.0696 (13.5), 337.0695 (66.1), 323.0903 (13.2), 313.0696 (89.3), 295.0590 (5.2), 283.0594 (100), 271.0592 (5.3), 243.0373 (0.3)	20	Isovitexin*	OA, OG, OK, OS, OTF	Flavonoid C-glycoside
5	9.5	$C_{21}H_{20}O_{10}$	433.1129	433.1129	0.02	415.1025 (80.6), 397.0908 (69.7), 379.0813 (21.2), 367.0806 (31), 355.0788 (5), 351.0879 (10), 343.0813 (13.2), 337.0709 (26), 323.0907 (9.5), 313.0707 (100), 295.0614 (9.5), 283.0600 (35.5), 271.0595 (3.3), 243.0284 (3.2), 217.0495 (2.8)	20	Vitexin*	OA, OG, OK, OS, OTF	Flavonoid-C-glycoside
6	9.7	$C_{27}H_{30}O_{15}$	595.1657	595.1657	0.09	465.5772 (0.5), 449.1077 (26.3), 433.1124 (2.6), 287.0563 (100), 273.0902 (0.5), 147.0633 (3.7), 129.0536 (4.6), 103.0412 (0.8), 85.0293 (3), 71.0515 (2.6)	8	Kaempferol-3-*O*-rutinoside*	OA, OB, OG, OK, OS, OTF	Flavonol glycoside

(*Continued*)

TABLE 2.1 (*Continued*) Compounds Identified from Leaf Extracts of *Ocimum* Species in Positive Ionization Mode by HPLC-QTOF-MS/MS

PEAK NO.	T_R *(MIN)*	*MOLECULAR FORMULA*	*CALC. M/Z [M+H]+*	*OBS. M/Z [M+H]+*	*ERROR (Δ PPM)*	*MS/MS DATA M/Z (% ABUNDANCE)*	*CE*	*IDENTIFICATION*	*DETECTION*	*CLASS*
7	9.9	$C_{21}H_{20}O_{12}$	465.1028	465.1028	–0.08	447.0964 (0.1), 432.1709 (0.04), 411.0693 (0.1), 370.0572 (0.1), 345.0567 (0.1), 328.0516 (0.03), 315.0436 (0.1), 303.0488 (100), 285.0313 (0.1), 257.0433 (0.1), 241.9372 (0.1), 229.0481 (0.1), 163.0609 (0.1), 145.0491 (1.1), 127.0384 (1.1), 109.0284 (0.1), 85.0287 (3), 61.0290 (0.4)	11	Quercetin-3-*O*-glucoside	OA, OB, OG, OK, OS, OTF	Flavonol glycoside
8	10.1	$C_{27}H_{30}O_{14}$	579.1708	579.1707	0.08	504.0224 (0.5), 433.1110 (80.8), 417.1135 (0.5), 399.1112 (0.7), 271.0591 (100), 225.0458 (0.6), 163.0387 (1), 129.0555 (2.5), 85.0275 (4.6), 71.0503 (1.8)	18	Apigenin-7-*O*-rutinoside	OG, OK, OS	Flavone glycoside
9	10.2	$C_{21}H_{18}O_{12}$	463.0871	463.0872	–0.16	422.1484 (0.2), 405.0572 (0.2), 387.0958 (0.2), 360.9824 (0.1), 301.0644 (0.8), 287.0539 (100), 257.1163 (0.2), 213.1611 (0.2), 187.0390 (0.1), 169.0525 (0.1), 151.0700 (0.2), 133.0252 (0.2)	15	Luteolin-7-*O*-glucuronide	OA, OB, OG, OK, OS, OTF	Flavone glycoside

(*Continued*)

TABLE 2.1 (*Continued*) Compounds Identified from Leaf Extracts of *Ocimum* Species in Positive Ionization Mode by HPLC-QTOF-MS/MS

PEAK NO.	T_R *(MIN)*	*MOLECULAR FORMULA*	*CALC. M/Z [M+H]*$^+$	*OBS. M/Z [M+H]*$^+$	*ERROR (Δ PPM)*	*MS/MS DATA M/Z (% ABUNDANCE)*	*CE*	*IDENTIFICATION*	*DETECTION*	*CLASS*
10	10.7	$C_{24}H_{22}O_{15}$	551.1032	551.1030	0.24	345.0640 (0.8), 303.0484 (100), 231.0511 (2), 181.0461 (0.4), 159.0251 (2.7), 145.0486 (3.4), 127.0376 (8.3), 109.0260 (5.1)	15	Quercetin 3-*O*-malonylglucoside	OA, OB, OG, OK, OS, OTF	Flavonol glycoside
11	11.2	$C_{21}H_{20}O_{11}$	449.1078	449.1077	0.3	413.0352 (0.8), 371.1098 (0.3), 351.4999 (0.4), 331.0887 (0.3), 303.4714 (0.3), 287.0536 (100), 267.1260 (0.4), 237.1223 (0.4), 209.1483 (0.5), 177.0495 (0.3), 149.0216 (1.2), 137.0545 (0.5), 127.1312 (0.2), 99.1156 (0.7), 85.1012 (0.7), 71.0852 (2.1), 57.0701 (1.5)	11	Luteolin-7-*O*-glucoside	OB, OG, OK, OS, OTF	Flavone glycoside
12	11.7	$C_{21}H_{20}O_{10}$	433.1129	433.1130	–0.05	301.0692 (1.3), 271.0590 (100), 187.1443 (0.3), 153.0133 (0.3)	15	Apigenin-7-*O*-glucoside	OA, OB, OG, OK, OS, OTF	Flavone glycoside
13	12	$C_{21}H_{18}O_{11}$	447.0922	447.0922	–0.02	287.0515 (0.1), 271.0591 (100), 229.0419 (0.1), 153.0175 (0.2), 119.0489 (0.1)	15	Apignin-7-*O*-glucuronide	OA, OB, OG, OK, OS, OTF	Flavone glycoside

(*Continued*)

TABLE 2.1 (*Continued*) Compounds Identified from Leaf Extracts of *Ocimum* Species in Positive Ionization Mode by HPLC-QTOF-MS/MS

PEAK NO.	T_R *(MIN)*	*MOLECULAR FORMULA*	*CALC. M/Z* $[M+H]^+$	*OBS. M/Z* $[M+H]^+$	*ERROR* (Δ *PPM)*	*MS/MS DATA M/Z (% ABUNDANCE)*	*CE*	*IDENTIFICATION*	*DETECTION*	*CLASS*
14	12.8	$C_{10}H_{12}O_2$	165.0910	165.0909	0.54	149.0597 (7), 137.0602 (68), 133.0646 (42), 124.0519 (100), 109.0290 (14), 105.0702 (62), 103.0548 (6), 79.0545 (6)	10	Eugenol*	OA, OB, OG, OK, OS, OTF	Propenyl phenol derivative
15	16.2	$C_{10}H_{10}O_3$	179.0703	179.0702	0.59	164.0465 (2.6), 151.0331 (3), 147.0431 (100), 136.0505 (9.3), 133.0640 (24.7), 123.0422 (3.9), 119.0486 (71.5), 115.0537 (9.2), 105.0698 (24.4)	15	Coniferaldehyde	OA, OB, OG, OK, OS, OTF	Propenyl phenol derivative
16	16.5	$C_{15}H_{10}O_6$	287.0550	287.0551	–0.35	269.0409 (4.5), 241.0485 (8.3), 213.0538 (5.9), 185.0587 (5.4), 161.0223 (16.2), 153.0180 (100), 135.0425 (31.1), 117.0347 (13.9), 89.0384 (19.5), 68.9974 (22.5)	40	Luteolin*	OA, OB, OG, OK, OS, OTF	Flavone

(*Continued*)

TABLE 2.1 (*Continued*) Compounds Identified from Leaf Extracts of *Ocimum* Species in Positive Ionization Mode by HPLC-QTOF-MS/MS

PEAK NO.	T_R *(MIN)*	*MOLECULAR FORMULA*	*CALC. M/Z [M+H]⁺*	*OBS. M/Z [M+H]⁺*	*ERROR (Δ PPM)*	*MS/MS DATA M/Z (% ABUNDANCE)*	*CE*	*IDENTIFICATION*	*DETECTION*	*CLASS*
17	16.9	$C_{15}H_{10}O_7$	303.0499	303.0500	–0.12	285.0410 (5), 274.0449 (3.8), 257.0471 (13.3), 239.0279 (6.5), 229.0481 (52.6), 219.0677 (4.3), 211.0357 (8.7), 201.0536 (28.4), 183.0428 (16.2), 173.0582 (17), 165.0170 (20.6), 153.0168 (100), 145.0639 (8), 137.0222 (50.8), 131.0490 (6.2), 121.0276 (14), 109.0282 (23.3), 95.0482 (6.1), 81.0349 (12.4), 68.9996 (25.5), 55.0189 (6.1)	40	Quercetin*	OA, OB, OG, OK, OS, OTF	Flavonol
18	18.8	$C_{17}H_{14}O_7$	331.0812	331.0812	0.12	316.0548 (10), 301.0325 (100), 298.0455 (27), 273.0368 (5.2), 242.0505 (1.9), 214.0607 (4.6), 198.0178 (0.8), 182.9914 (6.9), 154.9955 (2.2), 141.0896 (1.1), 125.0989 (0.8), 119.0489 (3.7), 105.0688 (0.8)	28	Isothymusin	OA, OB, OG, OK, OS, OTF	Flavone

(*Continued*)

TABLE 2.1 (*Continued*) Compounds Identified from Leaf Extracts of *Ocimum* Species in Positive Ionization Mode by HPLC-QTOF-MS/MS

PEAK NO.	*T_R (MIN)*	*MOLECULAR FORMULA*	*CALC. M/Z $[M+H]^+$*	*OBS. M/Z $[M+H]^+$*	*ERROR (Δ PPM)*	*MS/MS DATA M/Z (% ABUNDANCE)*	*CE*	*IDENTIFICATION*	*DETECTION*	*CLASS*
19	19	$C_{15}H_{10}O_5$	271.0601	271.0601	0.13	253.0472 (1.6), 243.0652 (4.7), 229.0493 (3.4), 197.0599 (2.9), 187.0393 (2.1), 169.0641 (2.6), 163.0384 (6.5), 153.0177 (100), 145.0280 (15.7), 131.0491 (3.2), 125.0239 (2.1), 119.0491 (46.8), 121.0286 (18.4), 111.0077 (3.8), 97.0292 (2.5), 91.0543 (42.6), 83.0142 (1.6), 68.9974 (23.6), 67.0184 (29.5)	40	Apigenin*	OA, OB, OG, OK, OS, OTF	Flavone
20	19.4	$C_{16}H_{12}O_6$	301.0707	301.0708	–0.29	286.0462 (78.1), 258.0512 (100), 229.0487 (6.4), 153.0168 (5.4)	33	Kaempferide	OA, OB, OG, OK, OS, OTF	Flavonol
21	19.5	$C_{15}H_{10}O_6$	287.0550	287.0553	–0.95	258.0531 (11.6), 241.0515 (5.1), 229.0480 (6.9), 213.0541 (20.3), 197.0585 (1.8), 185.0601 (8.4), 165.0175 (22), 153.0175 (100), 137.0231 (19.7), 121.0284 (58.3), 107.0491 (13.4), 93.0331 (17.3), 68.9972 (39.9),	40	Kaempferol*	OA, OB, OG, OK, OS, OTF	Flavonol

(*Continued*)

TABLE 2.1 (*Continued*) Compounds Identified from Leaf Extracts of *Ocimum* Species in Positive Ionization Mode by HPLC-QTOF-MS/MS

PEAK NO.	T_R *(MIN)*	*MOLECULAR FORMULA*	*CALC. M/Z* $[M+H]^+$	*OBS. M/Z* $[M+H]^+$	*ERROR* *(Δ PPM)*	*MS/MS DATA M/Z (% ABUNDANCE)*	*CE*	*IDENTIFICATION*	*DETECTION*	*CLASS*
22	19.6	$C_{17}H_{14}O_7$	331.0812	331.0811	0.3	316.0566 (78.13), 298.0456 (100), 286.0463 (3.52), 270.0504 (98.8), 242.0554 (8.2), 186.0668 (3.4), 137.0203 (9), 108.0191 (9.03	28	Cirsiliol	OA, OB, OG, OK, OS, OTF	Flavone
23	22.3	$C_{17}H_{14}O_6$	315.0863	315.0864	–0.3	300.0610 (14.4), 282.0510 (82.3), 271.0587 (3.8), 254.0559 (100), 226.0559 (7.7), 197.0572 (1.8), 170.0712 (1.8), 136.0142 (12.9), 119.0482 (6.5), 108.0189 (8.7)	28	Cirsimaritin	OA, OB, OG, OK, OS, OTF	Flavone
24	22.7	$C_{18}H_{16}O_7$	345.0969	345.0970	–0.27	330.0718 (53.6), 312.0612 (100), 297.0389 (2), 284.0662 (98.2), 269.0440 (8.9), 256.0716 (6.5), 241.0468 (2.6), 213.0543 (1.9), 185.0588 (1.7), 148.0501 (7.5), 136.0148 (4.4), 108.0203 (4.5)	28	Cirsilineol	OA, OB, OG, OK, OS, OTF	Flavone

(*Continued*)

TABLE 2.1 (*Continued*) Compounds Identified from Leaf Extracts of *Ocimum* Species in Positive Ionization Mode by HPLC-QTOF-MS/MS

PEAK NO.	T_R *(MIN)*	*MOLECULAR FORMULA*	*CALC. M/Z [M+H]*$^+$	*OBS. M/Z [M+H]*$^+$	*ERROR (Δ PPM)*	*MS/MS DATA M/Z (% ABUNDANCE)*	*CE*	*IDENTIFICATION*	*DETECTION*	*CLASS*
25	25.2	$C_{18}H_{16}O_7$	345.0969	345.0967	0.4	330.0713 (8.9), 315.0482 (100), 312.0612, (27), 300.0637 (0.71), 287.0530 (4.11), 256.0723 (1.31), 244.0643 (4.23), 228.0643 (0.24), 213.0376 (2.16), 197.0503 (0.68), 182.9919 (6.63), 175.0734 (0.33), 168.0064 (0.21), 154.9947 (2.29), 141.0677 (0.23), 133.0633 (4.7), 120.0145 (0.3), 10.0843 (0.4)	28	Nevadensin	OB, OG, OK, OS, OTF	Flavone
26	25.7	$C_{16}H_{12}O_5$	285.0757	285.0756	0.61	270.0494 (44.7), 242.0565 (100), 208.8208 (4.2), 153.0176 (6), 147.1118 (5), 133.0629 (7.4), 123.0769 (11.2), 105.0649 (8.8),	30	Acacetin	OA, OB, OG, OK, OS, OTF	Flavone
27	27	$C_{19}H_{18}O_7$	359.1125	359.1124	0.4	344.0870 (30), 326.0770 (89.4), 315.0856 (5), 298.0820 (100), 283.0574 (2), 270.0867 (3.9), 255.0625 (1.6), 239.0662 (0.7), 227.0672 (1.4), 211.0742 (1.2), 199.0725 (0.5), 183.0290 (1.7), 162.0658 (12.3), 147.0429 (1), 136.0145 (1.8), 108.0196 (3)	28	5-Desmethylsinensetin	OB, OG, OK, OS, OTF	Flavone

(*Continued*)

TABLE 2.1 (*Continued*) Compounds Identified from Leaf Extracts of *Ocimum* Species in Positive Ionization Mode by HPLC-QTOF-MS/MS

PEAK NO.	T_R *(MIN)*	*MOLECULAR FORMULA*	*CALC. M/Z* $[M+H]^+$	*OBS. M/Z* $[M+H]^+$	*ERROR (Δ PPM)*	*MS/MS DATA M/Z (% ABUNDANCE)*	*CE*	*IDENTIFICATION*	*DETECTION*	*CLASS*
28	31.3	$C_{11}H_{14}O_2$	179.1067	179.1068	−1.07	164.0818 (11), 151.0735 (21.3), 147.0806 (8.6), 138.0663 (100), 136.0526 (6.8), 133.0631 (3.5), 123.0442 (21.5), 107.0498 (9.7), 104.0617 (3.3)	15	Methyl eugenol*	OA, OB, OG, OK, OS, OTF	Propenyl phenol derivative
29	32.1	$C_{18}H_{16}O_6$	329.1020	329.1020	0.01	314.0765 (12.8), 296.0658 (69.8), 285.0731 (4.6), 268.0696 (100), 240.0777 (3.6), 205.9837 (2.6), 197.0435 (6), 182.0187 (3.9), 175.1437 (3.2), 169.0591 (2.1), 159.1133(1.8), 145.0316 (1.7), 136.0142 (7.8), 133.0647 (7.1), 119.0879 (2.3), 108.0181 (8.5), 105.0750 (2)	28	Salvigenin	OB, OK, OS, OTF	Flavone
30	33.3	$C_{11}H_{14}O_3$	195.1016	195.1016	0.16	180.0798 (3), 177.0902 (2), 167.0689 (5), 163.0746 (42), 154.0626 (100), 147.0422 (3), 139.0365 (16), 135.0790 (27), 133.0766 (5), 133.0502 (8), 107.0492 (21), 103.0540 (15), 91.0545 (4), 65.0347 (5)	13	Methoxyeugenol	OA, OB, OG, OK, OS, OTF	Propenyl phenol derivative

(*Continued*)

TABLE 2.1 (*Continued*) Compounds Identified from Leaf Extracts of *Ocimum* Species in Positive Ionization Mode by HPLC-QTOF-MS/MS

PEAK NO.	T_R *(MIN)*	*MOLECULAR FORMULA*	*CALC. M/Z* $[M+H]^+$	*OBS. M/Z* $[M+H]^+$	*ERROR* (Δ *PPM)*	*MS/MS DATA M/Z (% ABUNDANCE)*	*CE*	*IDENTIFICATION*	*DETECTION*	*CLASS*
31	36	$C_{19}H_{18}O_7$	359.1125	359.1125	0.08	344.0885 (5.5), 329.0641 (100), 311.0532 (36.6), 298.0817 (6.6), 286.0443 (1.8), 270.0853 (0.49), 255.0651 (0.4), 227.0539 (2.1), 214.0596 (0.3), 197.0071 (4.8), 178.9978 (0.6), 169.0116, 151.002 (0.6), 133.0635 (2.3), 113.0232 (0.5)	28	Gardenin B	OB, OG, OK, OS, OTF	Flavone
32	37.3	$C_{17}H_{14}O_5$	299.0914	299.0912	0.59	284.0667 (32), 256.0723 (100), 241.0476 (1.5), 227.0714 (1.5), 213.0558 (1.7), 197.0568 (1.6), 167.0332 (9.1), 133.0634 (3.4), 124.0155 (3.3)	30	Apigenin-7, 4′-dimethyl ether	OB, OG, OK, OS, OTF	Flavone
33	55.2	$C_{30}H_{48}O_3$	457.3676	457.3675	0.22	439.3551 (100), 425.3745 (6.4), 411.3604 (46.3), 393.3489 (6.6), 371.0990 (14), 338.3395 (8.2), 249.1829 (6.5), 191.1779 (17.8), 158.1526 (7.4)	25	Oleanolic acid	OA, OB, OG, OK, OS, OTF	Triterpenic acid

(*Continued*)

TABLE 2.1 (*Continued*) Compounds Identified from Leaf Extracts of *Ocimum* Species in Positive Ionization Mode by HPLC-QTOF-MS/MS

PEAK NO.	T_R *(MIN)*	*MOLECULAR FORMULA*	*CALC. M/Z* $[M+H]^+$	*OBS. M/Z* $[M+H]^+$	*ERROR (Δ PPM)*	*MS/MS DATA M/Z (% ABUNDANCE)*	*CE*	*IDENTIFICATION*	*DETECTION*	*CLASS*
34	57.9	$C_{30}H_{48}O_3$	457.3676	457.3679	–0.6	439.3553 (100), 411.3602 (39.3), 425.3745 (3), 393.3492 (5.7), 347.2470 (4.5), 338.3391 (5.8), 249.1829 (6.5), 203.1777 (4.2), 191.1777 (17.3), 174.1264 (7), 158.1524 (5.5)	25	Ursolic acid*	OA, OB, OG, OK, OS, OTF	Triterpenic acid
Compounds identified from leaf extracts of *Ocimum* species in negative ionization mode by HPLC-QTOF-MS/MS. (Reproduced from Pandey and Kumar (2016) with permission from Taylor & Francis)										
1	7	$C_{13}H_{16}O_{10}$	331.0671	331.0674	–0.98	283.2588 (11.6), 271.0439 (53.7), 211.0230 (57), 169.0130 (100), 151.00198 (17.7), 125.0230 (14.2)	24	Galloylglucose	OA, OB, OG, OK, OS, OTF	Phenolic acid ester
2	9	$C_7H_6O_5$	169.0142	169.0143	–0.41	161.0523 (0.2), 148.8981 (0.3), 135.0504 (0.4), 125.0258 (100), 107.0128 (1.5)	15	Gallic acid*	OA, OB, OG, OK, OS, OTF	Phenolic acid
3	9.8	$C_{16}H_{18}O_9$	353.0878	353.0880	–0.5	191.0559 (100), 179.0396 (0.7), 161.0254 (1.2), 135.0425 (0.6)	10	Chlorogenic acid*	OA, OB, OG, OK, OS, OTF	Phenolic acid
4	10.4	$C_9H_{10}O_4$	181.0506	181.0505	0.67	153.0176 (14.1), 108.0209 (100)	21	Ethyl protocatechuate	OA, OB, OG, OK, OS, OTF	Phenolic acid ester
5	10.7	$C_7H_6O_4$	153.0193	153.0193	0.02	143.0175 (3.1), 132.9078 (3.5), 123.0420 (6.1), 109.0280 (100), 103.0687 (2.4)	10	Protocatechuic acid*	OA, OB, OG, OK, OS, OTF	Phenolic acid

(*Continued*)

TABLE 2.1 (*Continued*) Compounds Identified from Leaf Extracts of *Ocimum* Species in Positive Ionization Mode by HPLC-QTOF-MS/MS

PEAK NO.	T_R *(MIN)*	*MOLECULAR FORMULA*	*CALC. M/Z* $[M+H]^+$	*OBS. M/Z* $[M+H]^+$	*ERROR* *(Δ PPM)*	*MS/MS DATA M/Z (% ABUNDANCE)*	*CE*	*IDENTIFICATION*	*DETECTION*	*CLASS*
6	13.1	$C_9H_8O_4$	179.0350	179.0350	–0.18	151.0020 (0.07), 147.0307 (0.08), 135.0450 (100), 132.0170 (0.05), 117.0342 (0.17), 107.0498 (0.55)	10	Caffeic acid*	OA, OB, OG, OK, OS, OTF	Phenolic acid
7	13.4	$C_9H_{10}O_5$	197.0455	197.0455	0.11	182.0191 (100), 166.9952 (61.3), 153.0558 (14.5), 138.0335 (23.9), 123.0067 (64.1), 106.0036 (8.1)	15	Syringic acid*	OA, OB, OG, OK, OS, OTF	Phenolic acid
8	13.6	$C_7H_8O_3$	137.0244	137.0243	0.98	119.4716 (13), 116.9957 (7), 108.0226 (89.1), 105.8005 (4.6), 100.9259 (23.3), 95.0127 (26.7), 93.0355 (100), 90.9054 (30.2), 82.7821 (4.8), 75.0285 (6.6), 69.0338 (13.1), 65.0062 (11.5)	30	*p*-Hydroxybenzoic acid	OA, OB, OG, OK, OS, OTF	Phenolic acid
9	15.1	$C_8H_8O_5$	183.0299	183.0299	–0.14	168.0054(3), 124.0151 (100)	18	Methyl gallate	OA, OB, OG, OK	Phenolic acid ester
10	15.8	$C_{11}H_{12}O_5$	223.0612	223.0611	0.58	208.0418 (100), 201.8391 (6.1), 193.0145 (55), 179.0685 (10.3), 164.0481 (63.7), 149.0252(27.7), 131.9295 (11.3), 108.8508 (2.8)	15	Sinapinic acid*	OA, OB, OG, OK, OS, OTF	Phenolic acid

(*Continued*)

TABLE 2.1 (*Continued*) Compounds Identified from Leaf Extracts of *Ocimum* Species in Positive Ionization Mode by HPLC-QTOF-MS/MS

PEAK NO.	T_R (MIN)	MOLECULAR FORMULA	CALC. M/Z $[M+H]^+$	OBS. M/Z $[M+H]^+$	ERROR (Δ PPM)	MS/MS DATA M/Z (% ABUNDANCE)	CE	IDENTIFICATION	DETECTION	CLASS
11	16.1	$C_9H_8O_3$	163.0401	163.0401	–0.05	149.0221 (1.5), 133.0297 (8.8), 121.0147 (3.1), 119.0461 (100), 116.9787 (1.4)	10	*p*-Coumaric acid*	OA, OB, OG, OK, OS, OTF	Phenolic acid
12	16.4	$C_{18}H_{16}O_8$	359.0772	359.0772	0.14	345.5925 (3.3), 312.9815 (2), 231.0497 (1.7), 223.0302 (4.2), 197.0485 (34.1), 179.0381 (20), 161.0268 (100), 150.9204 (1.5), 135.0470 (6.4), 123.0458 (3.6)	15	Rosmarinic acid*	OA, OB, OG, OK, OS, OTF	Phenolic acid
13	16.9	$C_8H_8O_4$	167.0350	167.0350	–0.32	152.0090 (1), 124.0130 (1), 119.0478 (1), 108.0193 (100)	15	Methyl protocatechuate	OA, OB, OG, OK, OS, OTF	Phenolic acid ester
14	17.2	$C_{10}H_{10}O_4$	193.0506	193.0506	0.03	178.0281 (23.9), 161.0816 (0.2), 149.0625 (4.1), 134.0391 (100), 121.0356 (0.9), 117.0294 (1), 111.0136 (0.4), 102.9498 (1.2)	15	Ferulic acid*	OA, OB, OG, OK, OS, OTF	Phenolic acid
15	21.5	$C_8H_8O_4$	167.0350	167.035	–0.37	152.0127 (19.5), 148.8912 (1), 137.0252 (1.4), 125.8712 (3.6), 108.0226 (100)	22	Vanillic acid*	OA, OB, OG, OK, OS, OTF	Phenolic acid
16	25.9	$C_{11}H_{12}O_4$	207.0663	207.0663	–0.3	179.0338 (23.4), 161.0233 (47.5), 135.0438 (100), 117.0370 (1), 107.0473 (2.2)	22	Ethyl caffeate	OA, OB, OG, OS, OTF	Phenolic acid ester

* = Compounds confirmed by reference standards; OA = *O. americanum;* OB = *O. basilicum;* OG = *O. gratissimum;* OK = *O. kilimandscharicum;* OS = *O. sanctum green;* OTF = *O. sanctum purple.*

Source: Reproduced from Pandey and Kumar (2016) with permission from Taylor & Francis.

ionization mode, although phenolic acids and their esters were analyzed in negative ionization mode because of higher sensitivity. The BPCs of the leaf extracts of six *Ocimum* species in positive and negative ionization modes are displayed in Figures 2.1 and 2.2, respectively.

2.4 IDENTIFICATION OF FLAVONOIDS

In this investigation, twenty-eight flavonoids were identified and characterized including four flavonoid *C*-glycosides, twelve flavones, five flavone *O*-glycosides, three flavonols and four flavonol *O*-glycosides.

2.5 FLAVONOID C-GLYCOSIDES

Peaks 1, 3, 4 and 5 were identified as flavonoid *C*-glycosides, isoorientin, orientin, isovitexin and vitexin, respectively, by co-chromatographic and mass spectral analysis with the authentic standards. Isoorientin (luteolin-6-*C*-glucoside) (peak 1) and orientin (luteolin-8-*C*-glucoside) (peak 3) showed $[M+H]^+$ ion at *m/z* 449.1078 and produced similar fragment ions in the CID-MS/MS scan, representing their isomeric structures. They produced similar base peak ions at *m/z* 329.0658 ($^{0,2}X^+$) by the loss of 120 Da from $[M+H]^+$ ion, which is characteristic of a hexose substitution in the aglycone moiety. They also generated fragment ion at *m/z* 431.0958 $[M+H-H_2O]^+$ (E_1^+), 413.0869 $[M+H-2H_2O]^+$ (E_2^+), 395.0764 $[M+H-3H_2O]^+$ (E_3^+), 383.0768 $[^{2,3}X^+-2H_2O]$, 353.0661 $[^{0,4}X^+-2H_2O]$ and 299.0538 $[M+H-150]^+$ ($^{0,1}X^+$). The relative intensities of the (E1$^+$) and ($^{0,1}X^+$) ions are 16% and 89% for isoorientin, whereas 36% and 44% for orientin, respectively, at a CID energy of 25 eV. These isomers were differentiated based on relative intensities of their characteristic ions.

Isovitexin (apigenin-6-*C*-glucoside) (peak 4) and vitexin (apigenin-8-*C*-glucoside) (peak 5) also have isomeric structures; they showed $[M+H]^+$ ion at *m/z* 433.1127. Isovitexin produced base peak ion at *m/z* 283.0594 $[M+H-150]^+$ ($^{0,1}X^+$), whereas vitexin at *m/z* 313.0707 $[M+H-120]^+$ ($^{0,2}X^+$). They also generated fragment ion at *m/z* 415.1025 $[M+H-H_2O]^+$ (E_1^+), 397.0908 $[M+H-2H_2O]^+$ (E_2^+), 379.0813 $[M+H-3H_2O]^+$ (E_3^+), 367.0806 $[^{2,3}X^+-2H_2O]$, 337.0709 $[^{0,4}X^+-2H_2O]$ and 295.0614 $[^{0,2}X^+-H_2O]$.

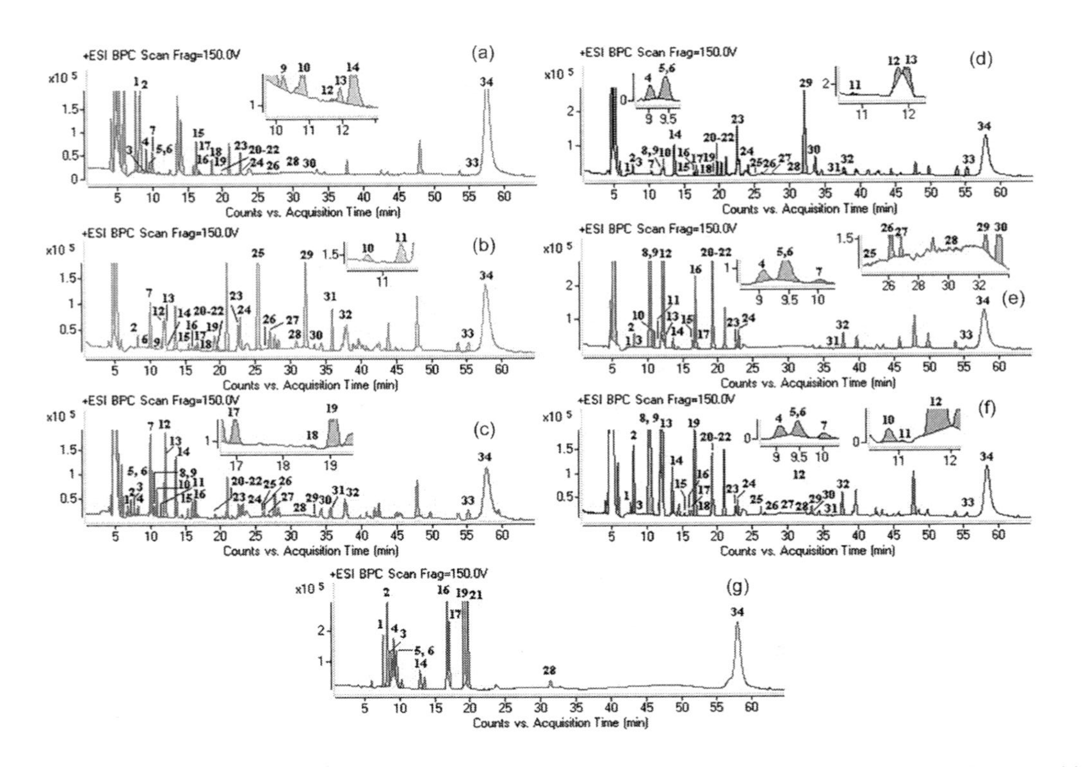

FIGURE 2.1 BPCs of leaf extract of (a) *O. americanum*, (b) *O. basilicum*, (c) *O. gratissimum*, (d) *O. kilimandscharicum*, (e) *O. sanctum* green, (f) *O. sanctum* purple and (g) mix reference standards in positive ionization mode. (Reproduced from Pandey and Kumar (2016) with permission from Taylor & Francis.)

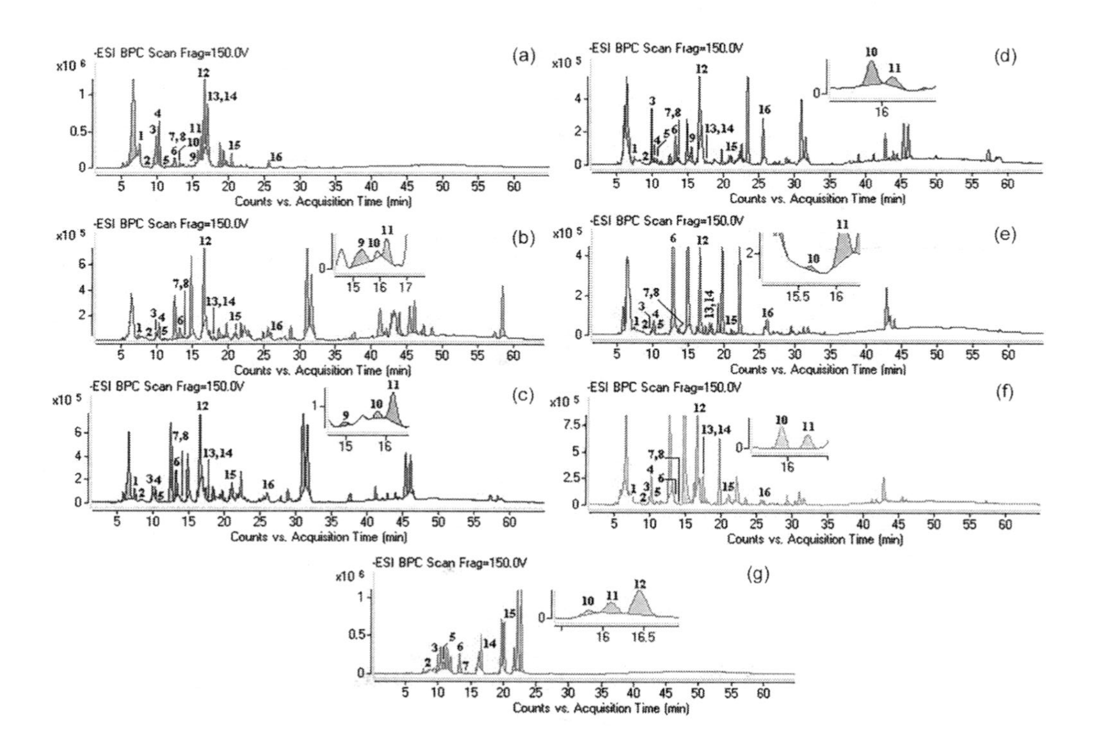

FIGURE 2.2 BPCs of leaf extract of (a) *O. americanum*, (b) *O. basilicum*, (c) *O. gratissimum*, (d) *O. kilimandscharicum*, (e) *O. sanctum* green, (f) *O. sanctum* purple and (g) mix reference standards in negative ionization mode. (Reproduced from Pandey and Kumar (2016) with permission from Taylor & Francis.)

2.6 FLAVONES AND THEIR *O*-GLYCOSIDES

Peaks 16, 18, 19, 22–27, 29, 31 and 32 were identified as luteolin, isothymusin, apigenin, cirsiliol, cirsimaritin, cirsilineol, nevadensin, acacetin, 5-desmethylsinensetin, salvigenin, gardenin B and apigenin-7, 4′-dimethyl ether, respectively, in which 16 and 19 were confirmed by comparison with authentic standards. All these are methoxylated flavones except luteolin and apigenin. In CID-MS/MS scan, luteolin (*m/z* 287.0550) (peak 16) and apigenin (*m/z* 271.0601) (peak 19) produced prominent RDA fragment ion at *m/z* 153.0180 ($^{1,3}A^+$ RDA) and other RDA fragment ions$^{1,3}B^+$, $^{0,4}B^+$ and $^{0,4}B^+$-H_2O. The $^{1,3}B^+$, $^{0,4}B^+$ and $^{0,4}B^+$-H_2O RDA fragment ions are specifically found for the flavones.

Peaks 18 (isothymusin) and 22 (cirsiliol) showed the same $[M+H]^+$ ion at *m/z* 331.0812 in the MS scan. In CID-MS/MS scan, they generated fragment ions at *m/z* 316.0548 $[M+H-15]^+$ corresponding to the loss of methyl radical and at *m/z* 298.0456 $[M+H-33]^+$ corresponding to the loss of methyl radical and a water molecule. The relative intensity of $[M+H-33]^+$ was 27% for isothymusin and 100% for cirsiliol, which has also been observed by Grayer et al. for differentiation of these isomeric flavones. Isothymusin (peak 18) also showed fragment ion at *m/z* 301.0325 $[M+H-30]^+$ corresponding to loss of two methyl groups and cirsiliol (peak 22) at *m/z* 270.0504 $[M+H-61]^+$ corresponding to loss of methyl radical, water and CO.

Peak 23 (cirsimaritin) showed $[M+H]^+$ ion at *m/z* 315.0864 and produced prominent ions at *m/z* 282.0510 $[M+H-33]^+$ and at *m/z* 254.0559 $[M+H-61]^+$ with less abundant ion at *m/z* 300.0610 $[M+H-15]^+$. Peaks 24 (cirsilineol) and 25 (nevadensin) showed the same $[M+H]^+$ ion at *m/z* 345.0970 indicating their isomeric structures. In CID-MS/MS scan, cirsilineol (peak 24) showed $[M+H-15]^+$ at *m/z* 330.0718, $[M+H-33]^+$ at *m/z* 312.0612 and $[M+H-61]^+$ at *m/z* 284.0662, whereas nevadensin (peak 25) showed $[M+H-15]^+$ at *m/z* 330.0713, $[M+H-30]^+$ at *m/z* 315.0482 and $[M+H-33]^+$ at *m/z* 312.0612. These fragmentations are in agreement with the reported literature (Grayer et al. 2001). Peaks 26 at *m/z* 285.0756 and 32 at *m/z* 299.0912 were identified as acacetin and apigenin-7, 4′-dimethyl ether, respectively. They are methyl derivatives of apigenin (peak 19). In CID-MS/MS scan, they showed initial loss of methyl radical produced fragment ion at *m/z* 270.0494 $[M+H-15]^+$ and 284.0667 $[M+H-15]^+$, respectively, and by subsequent loss of CO generated base peak ion at *m/z* 242.0565 $[M+H-43]^+$and 256.0723 $[M+H-43]^+$, respectively.

Peaks 27 (5-desmethylsinensetin) and 31 (gardenin B) also have isomeric structures, and they showed $[M+H]^+$ ion at *m/z* 359.1124. In the MS/MS analysis,

5-desmethylsinensetin (peak 27) generated fragment ion at *m/z* 344.0870 $[M+H-15]^+$, 326.0770 $[M+H-33]^+$ and 298.0820 $[M+H-61]^+$, whereas gardenin B (peak 31) generated fragment ion at *m/z* 344.0885 $[M+H-15]^+$, 329.0641 $[M+H-30]^+$, 311.0532 $[M+H-48]^+$ and 298.0817 $[M+H-61]^+$. The $[M+H-48]^+$ corresponding to loss of two methyl radicals and a water molecule. Peak 29 (salvigenin) showed $[M+H]^+$ ion at *m/z* 329.1020 and yielded fragment ions at *m/z* 314.0765 $[M+H-15]^+$, 296.0658 $[M+H-33]^+$ and 268.0696 $[M+H-61]^+$ as reported earlier.

Peaks 8, 9, 11, 12 and 13 were identified as apigenin-7-*O*-rutinoside (*m/z* 579.1707), luteolin-7-*O*-glucuronide (*m/z* 463.0872), luteolin-7-*O*-glucoside (*m/z* 449.1077), apigenin-7-*O*-glucoside (*m/z* 433.1130) and apigenin-7-*O*-glucuronide (*m/z* 447.0922). They produced the most prominent (Y_0^+) fragment ion at *m/z* 271.0591 and 287.0539 corresponding to aglycone unit owing to loss of sugar moiety (Cuyckens et al. 2000). Apigenin-7-*O*-rutinoside (peak 8) also generated fragment ion at *m/z* 433.1110 (Y_1^+), which corresponds to the loss of terminal rhamnose unit (146 Da), and at *m/z* 417.1135 (Y^*), which corresponds to the loss of an internal dehydrated glucose moiety (162 Da). The presence of (Y_1^+) and (Y^*) ions indicated that peak 8 is an *O*-diglycoside.

2.7 FLAVONOLS AND THEIR *O*-GLYCOSIDES

Peaks 17, 20 and 21 were identified as quercetin (*m/z* 303.0500), kaempferide (*m/z* 301.0708) and kaempferol (*m/z* 287.0553), respectively, in which peaks 17 and 21 were confirmed by comparison with authentic standards. Quercetin (peak 17) and kaempferol (peak 21) generated prominent fragment ion at *m/z* 153.0168 ($^{1,3}A^+$ RDA) and other RDA fragment ions $^{0,2}A^+$ and $^{0,2}B^+$, which are characteristic ions of flavonols. Kaempferide (peak 20) is 4′-*O*-methyl derivative of kaempferol (peak 21) and generated fragment ion at *m/z* 286.0462 by initial loss of methyl radical and at *m/z* 258.0512 by subsequent loss of CO. Kaempferide also showed characteristic fragment ion at *m/z* 153.0168 ($^{1,3}A^+$) by RDA.

Peaks 2, 6, 7 and 10 were identified as quercetin-3-*O*-rutinoside (*m/z* 611.1607), kaempferol-3-*O*-rutinoside (*m/z* 595.1657), quercetin-3-*O*-glucoside (*m/z* 465.1028) and quercetin-3-*O*-malonylglucoside (*m/z* 551.1030), respectively, in which peaks 2 and 6 were confirmed by the authentic standard. These flavonols *O*-glycosides produced the base peak (Y_0^+) ion at *m/z* 303.0485 and 287.0539 corresponding to aglycone unit

(Cuyckens et al. 2000). Quercetin-3-*O*-rutinoside (peak 2) and kaempferol-3-*O*-rutinoside (peak 6) are *O*-diglycoside, and they also generated their characteristic (Y_1^+) ion at *m/z* 465.1013 and 449.1077, (Y^*) ion at *m/z* 449.103 and 433.1124, respectively.

2.8 IDENTIFICATION OF PROPENYL PHENOL DERIVATIVES

Peaks 14, 15, 28 and 30 were identified as eugenol (*m/z* 165.0909), coniferaldehyde (*m/z* 179.0702), methyl eugenol (*m/z* 179.1068) and methoxy eugenol (*m/z* 195.1016), respectively, in which peaks 14 and 28 were confirmed by the authentic standard. In CID-MS/MS scan, eugenol (peak 14), methyl eugenol (peak 28) and methoxy eugenol (peak 30) produced base peak ion at *m/z* 124.0519, 138.0673 and 154.0626, respectively, due to loss of propenyl radical from their $[M+H]^+$ ions. Eugenol (peak 14) and methoxy eugenol (peak 30) showed fragment ions at *m/z* 133.0646 and 163.0746, respectively, due to loss of CH_3OH. Further loss of CH_2O from fragment ion at *m/z* 163.0746 generated fragment ion at *m/z* 133.0766, indicating peak 30 is a methoxy derivative of eugenol. Peak 15 (coniferaldehyde) produced base peak ion at *m/z* 147.0431 corresponding to $[M+H-CH_3OH]^+$ and by the subsequent loss of CO generated second abundant fragment ion at *m/z* 119.0486. Coniferaldehyde also yielded $[M+H-CH3]^+$ and $[M+H-CO]^+$ fragments.

2.9 IDENTIFICATION OF TRITERPENIC ACIDS

Two isomeric triterpenic acids, peaks 33 and 34, were identified and characterized as oleanolic acid and ursolic acid, respectively. These isomeric acids showed similar $[M+H]^+$ at *m/z* 457.3675 and fragment ions in MS/MS scan but eluted at different retention times, 55.2 and 57.9 min. The identity of ursolic acid (peak 34) at t_R 57.9 min was confirmed by the authentic standard. The protonated molecular ions of oleanolic acid (peak 33) and ursolic acid (peak 34) generated base peak ion at *m/z* 439.3551 due to loss of H_2O. They also produced fragment ions at *m/z* 411.3604 $[M+H-COOH]^+$ and 393.3489 $[M+H-COOH-H_2O]$ as reported earlier.

2.10 IDENTIFICATION OF PHENOLIC ACIDS AND THEIR ESTERS

Analysis of phenolic acids and their esters was carried out in negative ionization mode; eleven phenolic acids and five phenolic acid esters were identified and characterized. Peaks 2, 3, 5–8, 10–12, 14 and 15 were identified as gallic acid (*m/z* 169.0143), chlorogenic acid (*m/z* 353.0878), protocatechuic acid (*m/z* 153.0193), caffeic acid (*m/z* 179.0350), syringic acid (*m/z* 197.0455), *p*-hydroxybenzoic acid (*m/z* 137.0243), sinapinic acid (*m/z* 223.0611), *p*-coumaric acid (*m/z* 163.0401), rosmarinic acid (*m/z* 359.0772), ferulic acid (*m/z* 193.0506) and vanillic acid (*m/z* 167.0350), respectively. All phenolic acids were confirmed by authentic standard except *p*-hydroxybenzoic acid. In CID-MS/MS scan, gallic acid (peak 2), protocatechuic acid (peak 5), caffeic acid (peak 6), *p*-hydroxybenzoic acid (peak 8) and *p*-coumaric acid (peak 11) generated major fragment ion at *m/z* 125.0258, 109.0280, 135.0450, 93.0355 and 119.0461, respectively, corresponding to $[M\text{-}H\text{-}CO_2]^-$ (Hossain et al. 2010b; Fang et al. 2002). Chlorogenic acid (peak 3) is an ester of caffeic acid and quinic acid which generated base peak ion at *m/z* 191.0559 corresponds to deprotonated quinic acid by the cleavage of intact caffeoyl and quinic acid fragments (Hossain et al. 2010b; Fang et al. 2002). Rosmarinic acid (peak 12), also a caffeic acid ester, which generated major fragment ion at *m/z* 161.0268 due to loss of water from fragment ion at *m/z* 179.0381 $[M\text{-}H\text{-}C_9H_8O_4]^-$, corresponds to deprotonated caffeic acid moiety (Hossain et al. 2010b). Syringic acid (peak 7), sinapinic acid (peak 10), ferulic acid (peak 14) and vanillic acid (peak 15) are *O*-methylated phenolic acids; they showed prominent fragment ion at *m/z* 182.0191 $[M\text{-}H\text{-}CH_3]^-$, 208.0418 $[M\text{-}H\text{-}CH_3]^-$, 134.0391 $[M\text{-}H\text{-}CH_3\text{-}CO_2]^-$ and 108.0226 $[M\text{-}H\text{-}CH_3\text{-}CO_2]^-$, respectively.

Peak 1 was identified as galloylglucose (*m/z* 331.0674), generated fragment ions at *m/z* 271.0462 due to removal of $C_2H_4O_2$ moiety by cross-ring fragmentation of a glucose molecule, *m/z* 169.0117 due to loss of glucose moiety ($C_6H_{10}O_5$) and *m/z* 125.0231 due to decarboxylation of the gallic acid moiety (Fang et al. 2002). Peaks 4 and 13 were identified as ethyl protocatechuate (*m/z* 181.0505) and methyl protocatechuate (*m/z* 167.0350), respectively. Ethyl protocatechuate (peak 4) showed fragment ion at *m/z* 153.0176 $[M\text{-}H\text{-}C_2H_4]^-$, 108.0209 $[M\text{-}H\text{-}C_2H_4\text{-}COOH]^-$ and methyl protocatechuate (peak 13) at *m/z* 152.0090 $[M\text{-}H\text{-}CH_3]^-$ and *m/z* 108.0193 $[M\text{-}H\text{-}CH_3\text{-}CO_2]^-$. Peak 9 was identified as methyl gallate (*m/z* 183.0299) and produced fragment ions at *m/z* 168.0054 $[M\text{-}H\text{-}CH_3]^-$ and 124.0151 $[M\text{-}H\text{-}CH_3\text{-}CO_2]^-$ (Barreto et al. 2008). Peak 16 was identified as ethyl caffeate (*m/z* 207.0663) and showed base peak ion at *m/z* 135.0438 corresponding to $[M\text{-}H\text{-}C_2H_4\text{-}CO_2]^-$.

3 Quantitation of Bioactive Phytochemicals in *Ocimum* Species and Its Marketed Herbal Formulations Using UHPLC-MS

Ocimum sanctum L., with phenolic acids, flavonoids, propenyl phenols and terpenoids as active pharmacological constituents, is a prominent therapeutic herb. It is used as an ingredient in many herbal formulations. Therefore, development of an analytical method for simultaneous determination of bioactive compounds of *O. sanctum* is required for quality control of herbal formulations. The aim of this study was to develop and validate a new, rapid, sensitive and selective ultra high performance liquid chromatograhy electrospray tandem mass spectrometry (UHPLC-ESI-MS/MS) strategy for simultaneous determination of bioactive markers including phenolic acids, flavonoids, propenyl phenol and terpenoid in the leaf extract and marketed herbal formulations of *O. sanctum*. The simultaneous determination was carried out using negative electrospray ionization with multiple reactions monitoring (MRM) mode. Chromatographic separation was achieved on an Acquity UHPLC BEH C_{18} column using gradient elution with 0.1% formic acid in water and 0.1% formic acid in acetonitrile. Principal component analysis (PCA) was carried out to correlate and discriminate eight geographical collections of *O. sanctum* based on quantitative data of analytes. The developed method was validated as per International Conference on Harmonization (ICH, 2005) guidelines and found to be accurate with overall recovery in the range from 95.09% to 104.84% (RSD $\leq$ 1.85%), precise (RSD $\leq$ 1.98%) and linear ($r^2 \geq 0.9971$) over the concentration range of 0.5–1,000 ng/mL. Ursolic acid was found to be the most abundant marker in all the investigated samples except for the marketed tablet. The established method is simple, rapid and sensitive and hence could be reliably utilized for the quality control of *O. sanctum* and derived herbal formulations.

3.1 PLANT EXTRACTS AND MARKETED FORMULATIONS

The leaves of *O. sanctum* (Figure 3.1) were collected from various geographical regions of India such as Nauni, Solan (Himachal Pradesh), Lucknow (Uttar Pradesh), Jabalpur (Madhya Pradesh) and Kolkata (West Bengal). The voucher specimens of *O. sanctum* from Uttar Pradesh (U.P.), Madhya Pradesh (M. P.) and West Bengal (W. B.) have been deposited in the Botany Department of CSIR-Central Drug Research Institute, Lucknow, India. The voucher specimen of *O. sanctum* from Himachal Pradesh (H. P.) has been deposited in the Department of Forest Products, Dr. Yashwant Singh Parmar University of Horticulture and Forestry, Nauni, Solan, Himachal Pradesh, India. The details of plant samples of six *Ocimum* species are given in Section 2.1.

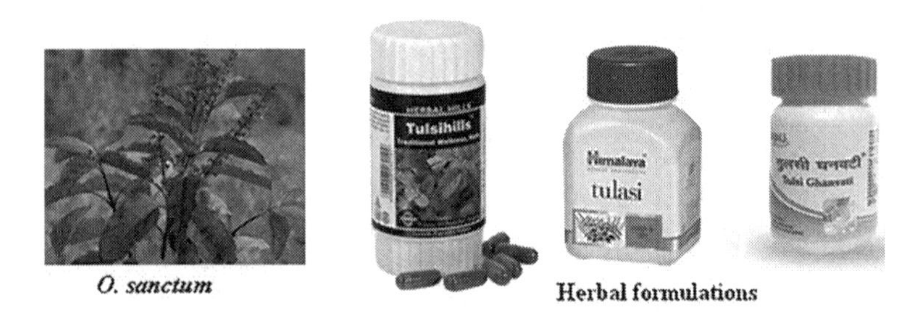

FIGURE 3.1 Pictures of *O. sanctum* and its herbal formulations.

3.2 INSTRUMENTATION AND ANALYTICAL CONDITIONS FOR QUANTITATION OF 15 ANALYTE IN SIX *OCIMUM* SPECIES

3.2.1 UHPLC-QqQ$_{LIT}$-MS/MS Conditions

The UHPLC-QqQ$_{LIT}$-MS/MS analysis was performed on a Waters Acquity UPLC™ system (Waters, Milford, MA, USA) interfaced with a hybrid linear ion trap triple-quadrupole mass spectrometer (API 4000 QTRAP™ MS/MS system from AB Sciex, Concord, ON, Canada) equipped with an electrospray (Turbo V) ion source. The Waters Acquity UPLC™ system was equipped with a binary solvent manager, sample manager, column oven and PDA. AB Sciex Analyst software version 1.5.1 was used to control the LC-MS/MS system and for data acquisition and processing. All the statistical calculations related to quantitative analysis were performed using Graph Pad Prism software version 5. The UHPLC separation of selected analytes and IS was achieved on an Acquity UPLC BEH C_{18} column (50 mm × 2.1 mm id, 1.7 μm) at a column temperature of 50°C. The analysis was completed with a gradient elution of 0.1% formic acid in water (A) and 0.1% formic acid in acetonitrile (B) as the mobile phase at a flow rate of 0.3 mL/min. The 13 min UHPLC gradient system was as follows: 0–2 min, 10%–10% B; 2–3 min, 10%–20% B; 3–4.5 min, 20%–20% B; 4.5–5.2 min, 20%–24% B; 5.2–6 min, 24%–24% B; 6–7.5 min, 24%–40% B; 7.5–7.8 min, 40%–50% B; 7.8–8.5, min 50%–70% B; 8.5–9 min, 70%–70% B; 9–9.3 min, 70%–95% B; 9.3–11.3 min, 95%–95% B; 11.3–12.3 min, 95%–10% B; and 12.3–13 min, 10%–10% B. The sample injection volume was 1 μL.

All the analytes and IS were detected using negative electrospray ionization with a precursor ion scan, and mass spectra were recorded in the range of *m/z* 100–1,000 at a cycle time of 9 s with a step size of 0.1 Da. Nitrogen was used as the nebulizer (GS1), heater (GS2) and curtain gas (CUR), as well as the collision activated dissociation (CAD) gas. Simultaneous quantitation of analytes was carried out using MRM acquisition mode at the unit resolution. The transitions and optimized compound dependent MRM parameters, declustering potential (DP), entrance potential (EP), collision energy (CE) and cell exit potential (CXP) for each analyte and IS, are listed in Table 3.1. The dwell time for all analytes was set at 200 ms. Optimized source parameters were as follows: ion spray voltage set at –4,200 V, curtain gas, nebulizer gas and heater gas set at 20, 20 and 20 psi, respectively, with a source temperature of 550°C. The CAD gas was set at medium, and the interface heater was on.

TABLE 3.1 MRM Compound Dependent Parameters of 15 Analytes and IS

PEAK NO.	T_R *(MIN)*	*ANALYTE*	*Q1 MASS (DA)*	*Q3 MASS (DA)*	*DP (V)*	*EP (V)*	*CE (EV)*	*CXP (V)*
1	0.65	Gallic acid	169	125	−59	−8	−21	−10
2	1.04	Protocatechuic acid	153	109	−64	−5	−22	−9
3	2.5	Chlorogenic acid	353	191	−60	−10	−30	−10
4	2.74	Caffeic acid	179	135	−48	−8	−21	−11
5	3.67	Ferulic acid	193	134	−58	−5	−23	−9
6	3.7	Rutin	609	301	−197	−10	−45	−17
7	3.76	Sinapinic acid	223	149	−55	−5	−27	−9
8	3.83	Kaempferol-3-*O*-rutinoside	593	285	−105	−5	−47	−8
9	4.29	Rosmarinic acid	359	161	−65	−10	−22	−8
10	5.55	Quercetin	301	151	−107	−9	−31	−12
11	5.59	Luteolin	285	133	−139	−10	−38	−12
12	5.82	Andrographolide (IS)	349	287	−80	−5	−16	−8
13	6.68	Apigenin	269	117	−71	−5	−45	−9
14	6.83	Kaempferol	285	239	−95	−5	−39	−15
15	7.74	Eugenol	163	148	−49	−5	−19	−13
16	10.07	Ursolic acid	455.1	455	−130	−10	−9	−13

Source: Reproduced from Pandey et al. (2016) with permission from Royal Society of Chemistry.

3.3 OPTIMIZATION OF UHPLC-QqQ_{LIT}-MS/MS CONDITIONS FOR QUANTITATIVE ANALYSIS OF 15 ANALYTES IN SIX *OCIMUM* SPECIES

For accomplishing optimal separation of all analytes in a short analysis time, UHPLC conditions such as the column, mobile phase, gradient elution program, flow rate, injection volume and column temperature were optimized. An Acquity UPLC BEH C_{18} column (50 mm × 2.1 mm id, 1.7 μm) was ultimately chosen for good separation efficiency and better peak shapes. Further optimization results showed that mobile phase system composed of 0.1% formic acid in water and 0.1% formic acid in acetonitrile achieved good separation along with better ionization at a flow rate of 0.3 mL/min and 50°C column temperature within 13 min.

The mass spectrometric conditions were optimized by direct infusion of 50 ng/mL solution of each analyte into mass spectrometer at a flow rate of 10 µL/min using a Harvard syringe pump (Harvard Apparatus, South Natick, MA, USA). MS spectra were recorded in both positive and negative ionization modes. Finally, negative ionization mode was selected due to the high signal sensitivity of all target analytes. Further to achieve most abundant, specific and stable MRM transition for each analyte, the compound dependent MRM parameters (DP, EP, CE and CXP) and source parameters (curtain gas, GS1, GS2 and ion source temperature) were optimized. The optimized UHPLC-MS/MS method in MRM acquisition mode was applied to quantify fifteen bioactive constituents in the six *Ocimum* species using andrographolide as an IS.

3.4 ANALYTICAL METHOD VALIDATION FOR QUANTITATIVE ANALYSIS OF 15 ANALYTES IN SIX *OCIMUM* SPECIES

This UHPLC-QqQ_{LIT}-MS/MS method for quantitation was validated according to the guidelines of International Conference on Harmonization (ICH, Q2R1, 2005) by determining linearity, LOD, LOQ, precision, accuracy and solution stability.

3.4.1 Linearity, LOD and LOQ

The stock solution was diluted with LC-MS grade acetonitrile to provide a series of concentrations in the range of 0.5–500 ng/mL for the construction of calibration curves. The linearity of calibration was determined using the analytes-to-IS peak area ratios versus the nominal concentration, and the calibration curves were constructed with a weight ($1/x^2$) factor by least-squares linear regression. The LOD and LOQ were determined based on calibration curve method by the following equations: LOD = $(3.3 \times S_{xy})/S_a$ and LOQ = $(10 \times S_{xy})/S_a$, where S_{xy} is the residual standard deviation of the regression line, and S_a is the slope of a calibration curve. All the calibration curves indicated good linearity with correlation coefficients (r^2) from 0.9989 to 1.0000 within the test ranges. The LOD for 15 analytes varied from 0.041 to 0.357 ng/mL and LOQ from 0.124 to 1.082 ng/mL.

3.4.2 Precision, Stability and Recovery

The intra- and inter-day variations were checked to determine the precision of the developed method. It was completed by determining fifteen analytes with IS in six replicates during a single day and by duplicating the experiments on three consecutive days. Variations of the peak area were taken as the measures of precision and expressed as % RSD. The overall intra- and inter-day precisions were not more than 1.95%. Stability of sample solutions stored at room temperature was examined by replicate injections of the sample solution at 0, 2, 4, 8, 12 and 24 h. The RSD values of stability of the 15 analytes were ≤2.91%.

A recovery test was carried out to evaluate the accuracy of this method. The test was performed by adding known amounts of the 15 analytical standards at low, medium and high levels into samples. The spiked samples were analyzed at each level by the proposed method in triplicate, and average recoveries were determined. The analytical method developed had good accuracy with overall recovery in the range from 95.10% to 103.04% (RSD ≤ 1.68%).

3.5 QUANTITATIVE ANALYSIS OF 15 ANALYTES IN SIX *OCIMUM* SPECIES SAMPLES

The developed UHPLC-QqQ_{LIT}-MS/MS method was successfully applied for the simultaneous quantitative determination of fifteen bioactive constituents in the leaf extracts of six *Ocimum* species. The content of 15 bioactive constituents is summarized in Table 3.2. Quantitative analysis indicated that ursolic acid with the content range of 0.373–14.1 mg/g and rosmarinic acid with the content range of 1.653–10.966 mg/g were the major constituents in all the analyzed *Ocimum* species except for *O. americanum*. Rosmarinic acid (10.966 mg/g) and rutin (10.9 mg/g) were the predominant constituents in *O. americanum*.

The total contents of seven phenolic acids were found highest (14.182 mg/g) in *O. americanum* and lowest (4.7 mg/g) in *O. sanctum* green; similarly, total contents of seven flavonoids were found highest (12.42 mg/g) in *O. americanum* and lowest (2.062 mg/g) in *O. killimandscharicum*. The content of terpenoid

TABLE 3.2 The Content (mg/g) of Fifteen Bioactive Constituents in the Leaf Extracts of Six *Ocimum* Species

	LEAF EXTRACTS					
ANALYTES (μG/G)	*O. americanum*	*O. basilicum*	*O. gratissimum*	*O. killimandscharicum*	*O. sanctum GREEN*	*O. sanctum PURPLE*
Phenolic acids						
Gallic acid	0.397	0.255	0.367	0.221	0.282	0.316
Protocatechuic acid	0.730	0.610	0.587	0.683	0.593	0.830
Chlorogenic acid	0.460	0.180	0.803	0.463	1.113	0.321
Caffeic acid	1.080	0.920	2.433	1.413	0.390	1.007
Ferulic acid	0.337	0.547	0.447	0.315	0.447	0.357
Sinapinic acid	0.212	0.239	0.237	0.254	0.222	0.233
Rosmarinic acid	10.967	7.600	7.867	8.067	1.653	8.000
Total	14.182	10.351	12.740	11.416	4.701	11.063

(*Continued*)

TABLE 3.2 (*Continued*) The Content (mg/g) of Fifteen Bioactive Constituents in the Leaf Extracts of Six *Ocimum* Species

ANALYTES	*LEAF EXTRACTS*					
(μG/G)	*O. americanum*	*O. basilicum*	*O. gratissimum*	*O. killimandscharicum*	*O. sanctum GREEN*	*O. sanctum PURPLE*
Flavonoids						
Rutin	10.900	1.653	0.920	0.693	0.074	0.173
Kaempferol-3-*O*-rutinoside	0.370	0.297	0.322	0.252	0.207	0.222
Quercetin	0.039	0.047	0.068	0.044	0.040	0.041
Luteolin	0.643	0.683	0.643	0.697	1.117	1.530
Apigenin	0.095	0.134	0.123	0.119	0.700	0.443
Kaempferol	0.373	0.301	0.326	0.257	0.211	0.226
Total	12.421	3.116	2.402	2.062	2.348	2.636
Propenyl phenol						
Eugenol	0.146	0.034	0.045	0.207	0.094	0.043
Terpenoid						
Ursolic acid	0.3734	8.0333	6.9333	14.1	1.4733	4.8

Source: Reproduced from Pandey et al. (2016) with permission from Royal Society of Chemistry.

(ursolic acid) was found highest (14.1 mg/g) in *O. killimandscharicum* and lowest (0.373 mg/g) in *O. americanum*; similarly, the content of propenyl phenol (eugenol) was found highest (0.206 mg/g) in *O. killimandscharicum* and lowest (0.034 mg/g) in *O. basilicum*.

Results of the overall quantitative analysis indicated that *O. killimandscharicum* contained the maximum amount of ursolic acid and eugenol, the major bioactive constituents, and showed the highest total content (28.090 mg/g) of 15 bioactive constituents compared to other samples. *O. killimandscharicum* is the less explored species of *Ocimum* genus. This information could be helpful for better swapping of *Ocimum* species.

3.6 INSTRUMENTATION AND ANALYTICAL CONDITIONS FOR QUANTITATION OF 23 ANALYTE IN GEOGRAPHICAL SAMPLES OF *O. SANCTUM* AND THEIR HERBAL FORMULATIONS

The UHPLC-ESI-MS/MS analysis was performed on a Waters Acquity UPLC™ system (Waters, Milford, MA, USA) interfaced with a hybrid linear ion trap triple-quadrupole mass spectrometer (API 4000 QTRAP™ MS/MS system from AB Sciex, Concord, ON, Canada) equipped with an electrospray (Turbo V) ion source. The Waters Acquity UPLC™ system was equipped with a binary solvent manager, sample manager, column oven and PDA. AB Sciex Analyst software version 1.5.1 was used to control the LC-MS/MS system and for data acquisition and processing. All the statistical calculations related to quantitative analysis were performed on Graph Pad Prism software version 5.

3.6.1 UHPLC Conditions

The chromatographic separation of selected analytes and ISs was achieved on an Acquity UPLC BEH C_{18} column (50 mm × 2.1 mm, 1.7 μm; Waters, Milford, MA, USA) at a column temperature of 50°C. The analysis was completed with gradient elution of 0.1% formic acid in water (A) and 0.1% formic acid in acetonitrile (B) as the mobile phase. The 13 min UPLC gradient system was as follows: 0–2 min, 5%–5% B; 2–3 min, 5%–20% B; 3–4.5 min, 20%–20% B;

4.5–5.2 min, 20%–24% B; 5.2–6 min, 24%–24% B; 6–7.5 min, 24%–40% B; 7.5–7.8 min, 40%–50% B; 7.8–8.5 min, 50%–70% B; 8.5–9 min, 70%–70% B; 9–9.3 min, 70%–90% B; 9.3–11.3 min, 90%–90% B; 11.3–12.3 min, 90%–5% B; and 12.3–13 min, 5%–5% B. Sharp and symmetrical peaks were obtained at a flow rate of 0.3 mL/min with a sample injection volume of 1 μL.

3.6.2 MS Conditions

A precursor ion scan was used for the screening and MRM acquisition mode for the quantitation of 23 bioactive markers in *O. sanctum*. All the analytes with an IS were detected using negative electrospray ionization with precursor ion scan, and mass spectra were recorded by scanning in the range of *m/z* 100–1000 at a cycle time of 9 s with a step size of 0.1 Da. Nitrogen was used as the nebulizer, heater and curtain gas, as well as the CAD gas. Optimized source parameters were as follows: ion spray voltage set at −4,200 V, curtain gas, nebulizer gas (GS1) and heater gas (GS2) set at 20, 20 and 20 psi, respectively, and source temperature set at 550°C. The CAD gas was set at medium, and the interface heater was on.

Simultaneous quantitation of analytes was carried out using MRM acquisition mode at the unit resolution, and its conditions were optimized for each compound during infusion. The transitions and optimized compound-dependent MRM parameters: DP, EP, CE and CXP for each analyte and IS are listed in Table 3.3.

TABLE 3.3 MRM Compound Dependent Parameters of 23 Analytes and IS

PEAK NO.	T_R *(MIN)*	*ANALYTE*	*Q1 MASS (DA)*	*Q3 MASS (DA)*	*DP (V)*	*EP (V)*	*CE (EV)*	*CXP (V)*
1	0.67	Gallic acid	169	125	−59	−8	−21	−10
2	1.08	Protocatechuic acid	153	109	−64	−5	−22	−9
3	2.14	Epicatechin	289	203	−120	−9	−29	−6
4	2.21	Catechin	289	203	−110	−10	−29	−8
5	2.55	Chlorogenic acid	353	191	−60	−10	−30	−10
6	2.79	Caffeic acid	179	135	−48	−8	−21	−11
7	3.46	Quercetin-3,4′-diglucoside	625	463	−97	−6	−25	−15
8	3.62	Ellagic acid	300.9	145	−77	−5	−55	−10
9	3.7	Ferulic acid	193	134	−58	−5	−23	−9

(Continued)

TABLE 3.3 (*Continued*) MRM Compound Dependent Parameters of 23 Analytes and IS

PEAK NO.	T_R *(MIN)*	*ANALYTE*	*Q1 MASS (DA)*	*Q3 MASS (DA)*	*DP (V)*	*EP (V)*	*CE (EV)*	*CXP (V)*
11	3.8	Sinapinic acid	223	149	−55	−5	−27	−9
12	3.87	Kaempferol-3-*O*-rutinoside	593	285	−105	−5	−47	−8
13	4.33	Rosmarinic acid	359	161	−65	−10	−22	−8
14	4.38	Vanillic acid	167	108	−55	−6	−22	−9
15	4.69	Scutellarein	285	117	−95	−5	−48	−10
16	5.59	Quercetin	301	151	−107	−9	−31	−12
17	5.62	Luteolin	285	133	−139	−10	−38	−12
18	5.66	Quercetin dihydrate	337	301	−43	−9	−13	−26
19	5.83	Andrographolide (IS)	349	287	−80	−5	−16	−8
20	6.73	Apigenin	269	117	−71	−5	−45	−9
21	6.83	Kaempferol	285	239	−95	−5	−39	−15
22	7.76	Eugenol	163	148	−49	−5	−19	−13
23	8.38	Chrysin	253	143	−75	−8	−41	−7
24	10.1	Ursolic acid	455.1	455	−130	−10	−9	−13

Source: Reproduced from Pandey et al. (2015) with permission from John Wiley and Sons.

3.7 OPTIMIZATION OF UHPLC-QqQ_{LIT}-MS/MS CONDITIONS FOR QUANTITATIVE ANALYSIS OF 23 ANALYTES IN *O. SANCTUM* SAMPLE AND THEIR HERBAL FORMULATIONS

Each investigated analyte was infused into the mass spectrometer by flow injection analysis (FIA), and the precursor ion and, at least, two product ions were preliminarily selected in both positive ion and negative ion modes. Due to the carboxyl group in the chemical structure of acids and the phenolic hydroxyl group in the chemical structure of flavonoids and propenyl phenol, all compounds exhibited good signal sensitivity in negative ionization mode. The compound dependent MRM parameters: DP, EP, CE and CXP were optimized by injecting the individual standard solution into the mass spectrometer to achieve the most abundant, specific and stable MRM transition for each investigated compound. The source parameters including the curtain gas, GS1, GS2 and ion source temperature were further optimized to get the highest abundance of precursor-to-product ions. The optimized compound dependent parameters and source parameters were combined, and finally, the optimized UHPLC-ESI-MS/MS method in MRM acquisition mode was applied to quantify 23 bioactive markers in the *O. sanctum* samples using andrographolide as an IS.

The MS spectra generated for all the compounds by ESI-MS in the negative ion mode gave the deprotonated molecule $[M\text{-}H]^-$. Nine phenolic acids, peaks 1, 2, 5, 6, 8, 9, 11, 13 and 14, were identified as gallic acid, protocatechuic acid, chlorogenic acid, caffeic acid, ellagic acid, ferulic acid, sinapinic acid, rosmarinic acid and vanillic acid, respectively. The major fragment ion (base peak) in the MS/MS spectra of the $[M\text{-}H]^-$ ions of gallic acid, *m/z* 169 $[M\text{-}H]^-$, protocatechuic acid, *m/z* 153 $[M\text{-}H]^-$ and caffeic acid, *m/z* 179 $[M\text{-}H]^-$ was due to the loss of a CO_2 molecule, providing an anion of $[M\text{-}H\text{-}CO_2]^-$ (Fang et al. 2002). Chlorogenic acid, *m/z* 353 $[M\text{-}H]^-$, generated the major fragment ion at *m/z* 191, corresponds to deprotonated quinic acid by the cleavage of intact caffeoyl and quinic acid fragments (Hossain et al. 2010b) and ellagic acid, *m/z* 300.9 $[M\text{-}H]^-$, at *m/z* 145 due to loss of CO from a fragment ion at *m/z* 173 (Gu et al. 2013; Ying et al. 2013). Ferulic acid, *m/z* 193 $[M\text{-}H]^-$ and sinapinic acid, *m/z* 223 $[M\text{-}H]^-$ generated the major fragment ions at *m/z* 134 and *m/z* 149, respectively, corresponding to $[M\text{-}H\text{-}COO\text{-}CH3]^-$ (Sun et al. 2007). Rosmarinic acid, *m/z* 359 $[M\text{-}H]^-$, generated the major fragment ion at *m/z* 161 due to the loss of water from a fragment ion at *m/z* 179 ($[M\text{-}H\text{-caffeic acid}]^-$) (Hossain et al. 2010b) and vanillic acid, *m/z* 167 $[M\text{-}H]^-$, at *m/z* 108 due to the loss of methyl radicals followed by the loss of a CO_2 molecule (Fang et al. 2002). Twelve flavonoids including flavans (epicatechin and catechin), flavonol glycosides (quercetin-3, 4′-diglucoside, rutin and kaempferol-3-*O*-rutinoside), flavones (scutellarein, luteolin, apigenin and chrysin) and flavonols (quercetin, kaempferol and quercetin dihydrate) were identified and quantified in this study. Peaks 3 and 4 were identified as epicatechin *m/z* 289 $[M\text{-}H]^-$ and catechin, *m/z* 289 $[M\text{-}H]^-$, respectively, both yielded fragment ions at *m/z* 203 due to the loss of C_2H_2O from a fragment ion at *m/z* 245 ($[M\text{-}H\text{-}CH_2CHOH]^-$). Peaks 7, 10 and 12 were identified as quercetin-3, 4′-diglucoside, *m/z* 625 $[M\text{-}H]^-$, rutin, *m/z* 609 $[M\text{-}H]^-$, and kaempferol-3-*O*-rutinoside, *m/z* 593 $[M\text{-}H]^-$, respectively, yielded fragment ions at *m/z* 463, *m/z* 301 and *m/z* 285, respectively, due to *O*-glycosidic cleavage. Peaks 15, 17, 20 and 23 were identified as scutellarein, *m/z* 285 $[M\text{-}H]^-$, luteolin, *m/z* 285 $[M\text{-}H]^-$, apigenin, *m/z* 269 $[M\text{-}H]^-$and chrysin, *m/z* 253 $[M\text{-}H]^-$, respectively. Scutellarein, luteolin and apigenin generated major fragment ions at *m/z* 117, *m/z* 133 and *m/z* 117, respectively, by retro-RDA reaction, whereas chrysin *m/z* 253 $[M-H]^-$generated a major fragment ion at *m/z* 143 due to the loss of C_3O_2 followed by C_2H_2O (Fabre et al. 2001). Peaks 16, 18 and 21 were identified as quercetin, *m/z* 301 $[M-H]^-$, quercetin dihydrate, *m/z* 337 $[M-H]^-$ and kaempferol, *m/z* 285 $[M-H]^-$, respectively. Quercetin yielded a major fragment ion at *m/z* 151 due to RDA reaction, quercetin dihydrate yielded a major fragment ion at *m/z* 301 due to loss of two water molecules and kaempferol yielded a fragment ion at *m/z* 239 corresponding to $[M-H_2O-CO]^-$.

Peak 22 was identified as eugenol (propenyl phenol), *m/z* 163 $[M-H]^-$, yielded a major fragment ion at *m/z* 148 due to the loss of methyl radical (Ying et al. 2013). Peak 24 was identified as ursolic acid (terpenoid), *m/z* 455.1 $[M-H]^-$. Routine MRM for ursolic acid quantitation with different parent and product ion was not suitable because it could not have collided into fragments when CE was lower than 40 eV, or no dominant product ions were detected if CE was higher than 40 eV (Xia et al. 2011). Therefore in the MRM mode of this experiment, the CE in Q2 was set to a low value (9 eV) to minimize fragmentation, and the parent ion isolated in Q1 (455.1) passed through Q2 without fragmentation. In Q3 (455), the same ion was monitored. The IS andrographolide (peak 19), *m/z* 349 $[M-H]^-$, yielded a fragment ion at *m/z* 287 due to the loss of water followed by the loss of carbon dioxide.

3.8 ANALYTICAL METHOD VALIDATION FOR QUANTITATIVE ANALYSIS OF 23 ANALYTES IN *O. SANCTUM* SAMPLE AND THEIR HERBAL FORMULATIONS

This UHPLC-QqQ_{LIT}-MS/MS method for quantitation of 23 analytes was validated according to ICH (2005) guidelines.

3.8.1 Linearity, LOD and LOQ

The IS method was employed to calculate the content of 23 bioactive markers in *O. sanctum*. The stock solution was diluted with acetonitrile to ten different concentrations for the construction of calibration curves. The linearity of calibration was determined using the analytes-to-IS peak area ratios versus the nominal concentration, and the calibration curves were constructed with a weight ($1/x^2$) factor by least squares linear regression. The applied calibration model for all curves was $y = ax + b$, where y = peak area ratio (analyte/IS), x = concentration of the analyte, a = slope of the curve and b = intercept. The LODs and LOQs were measured with S/N of 3 and 10, respectively, as criteria. All the calibration curves indicated good linearity with correlation coefficients (r^2) from 0.9971 to 1.0000 within the test ranges. The LOD for 23 analytes varied from 0.041 to 0.357 ng/mL and LOQ from 0.124 to 1.082 ng/mL.

3.8.2 Precision, Stability and Recovery

The intra- and inter-day variations, which were chosen to determine the precision of the developed method, were investigated by determining twenty three analytes with IS in six replicates during a single day and by duplicating the experiments on three consecutive days. Variations of the peak area were taken as the measures of precision and expressed as % RSD. The overall intra- and inter-day precisions were not more than 1.98%. Stability of sample solutions stored at room temperature was investigated by replicate injections of the sample solution at 0, 2, 4, 8, 12 and 24 h. The RSD values of stability of the 23 analytes are ≤2.91%.

A recovery test was applied to evaluate the accuracy of this method. Three different concentration levels (high, middle and low) of the analytical standards were added to the samples in triplicate, and average recoveries were determined. The developed analytical method had good accuracy with overall recovery in the range from 95.10% to 103.04% (RSD ≤ 1.68%).

3.9 QUANTITATIVE ANALYSIS OF 23 ANALYTES IN *O. SANCTUM* SAMPLE AND THEIR HERBAL FORMULATIONS

The developed UHPLC-ESI-MS/MS method was applied to determine the content of twenty three markers in the eight leaf extracts and three marketed herbal formulations of *O. sanctum*. The content of 23 bioactive markers is summarized in Table 3.4, and graphical representations of results are shown in Figure 3.2 where it can be observed that the concentration of the 23 bioactive markers varied significantly in the leaf extracts collected from the different geographical regions of India and also in the different marketed herbal formulations of *O. sanctum*. Quantitative analysis showed that rutin was found below the detection level in the leaf extract from M. P. (OS-5); similarly, epicatechin

TABLE 3.4 The Content (mg/g) of 23 Markers in the Leaf Extracts and Marketed Herbal Formulations of *O. sanctum*

ANALYTES (MG/G)	*LEAF EXTRACTS*								*HERBAL FORMULATIONS*		
	OS-1	OS-2	OS-3	OS-4	OS-5	OS-6	OS-7	OS-8	HHOS	HMOS	PJOS
Phenolic acids											
Gallic acid	0.316	0.690	0.943	0.787	0.253	0.309	0.267	0.298	0.271	0.089	0.077
Protocatechuic acid	0.830	1.583	1.163	0.820	0.523	1.133	0.627	0.547	0.232	0.372	0.180
Chlorogenic acid	0.321	0.308	0.308	0.322	0.350	0.308	0.312	0.313	0.120	0.147	0.105
Caffeic acid	1.007	1.673	0.960	0.673	0.153	0.970	0.140	0.294	0.148	0.074	0.116
Ellagic acid	0.109	0.282	0.830	0.473	0.115	0.173	0.200	0.116	0.064	0.039	0.043
Ferulic acid	0.357	4.367	1.457	1.763	0.357	0.423	2.437	0.517	0.125	0.088	0.056
Sinapinic acid	0.233	0.230	0.221	0.272	0.235	0.397	0.318	0.210	0.078	0.075	0.080
Rosmarinic acid	8.000	6.533	2.897	2.193	0.227	1.343	0.241	0.228	2.940	2.180	1.723
Vanillic acid	0.107	0.116	0.107	0.109	0.115	0.107	0.110	0.054	0.035	0.032	0.032
Total	11.28	15.78	8.89	7.41	2.33	5.16	4.65	2.58	4.01	3.10	2.41
Flavonoids											
Epicatechin	Bdl	Bdl	0.085	Bdl	0.080	Bdl	Bdl	0.090	0.038	0.019	0.016
Catechin	Bdl	Bdl	0.051	Bdl	0.046	Bdl	Bdl	0.056	0.024	0.012	0.009
Quercetin-3,4′-diglucoside	0.304	0.301	0.326	0.300	0.299	0.297	0.303	0.302	0.092	0.094	0.090
Rutin	0.298	0.167	0.149	0.095	Bdl	0.155	0.220	0.142	0.062	0.050	0.030

(*Continued*)

TABLE 3.4 (*Continued*) The Content (mg/g) of 23 Markers in the Leaf Extracts and Marketed Herbal Formulations of *O. sanctum*

ANALYTES (MG/G)				*LEAF EXTRACTS*						*HERBAL FORMULATIONS*	
Kaempferol-3-*O*-rutinoside	0.222	0.203	0.205	0.206	0.204	0.199	0.211	0.199	0.074	0.073	0.064
Scutellarein	0.053	0.030	0.057	0.026	0.051	0.059	0.094	0.052	Bdl	0.006	Bdl
Quercetin	0.041	0.049	0.045	0.042	0.040	0.044	0.045	0.039	0.012	0.015	0.012
Luteolin	1.530	1.040	1.460	0.860	0.710	2.713	0.703	0.620	0.263	0.369	0.207
Quercetin dehydrate	0.171	0.175	0.165	0.182	0.178	0.216	0.177	0.167	0.052	0.057	0.053
Apigenin	0.443	0.200	1.220	0.943	1.460	1.370	0.245	0.124	0.069	0.312	0.050
Kaempferol	0.226	0.207	0.209	0.206	0.208	0.204	0.215	0.204	0.077	0.074	0.065
Chrysin	0.087	0.155	0.106	0.104	0.118	0.076	0.119	0.145	0.012	0.009	0.008
Total	3.37	2.53	4.08	2.96	3.39	5.33	2.33	2.14	0.78	1.09	0.60
Propenyl phenol											
Eugenol	0.046	0.061	0.129	0.186	0.101	0.178	0.156	0.110	0.016	0.060	0.080
Terpenoid											
Ursolic acid	5.050	16.133	13.000	12.733	7.533	7.233	2.100	3.250	4.640	19.100	0.623

Source: Reproduced from Pandey et al. (2015) with permission from John Wiley and Sons.

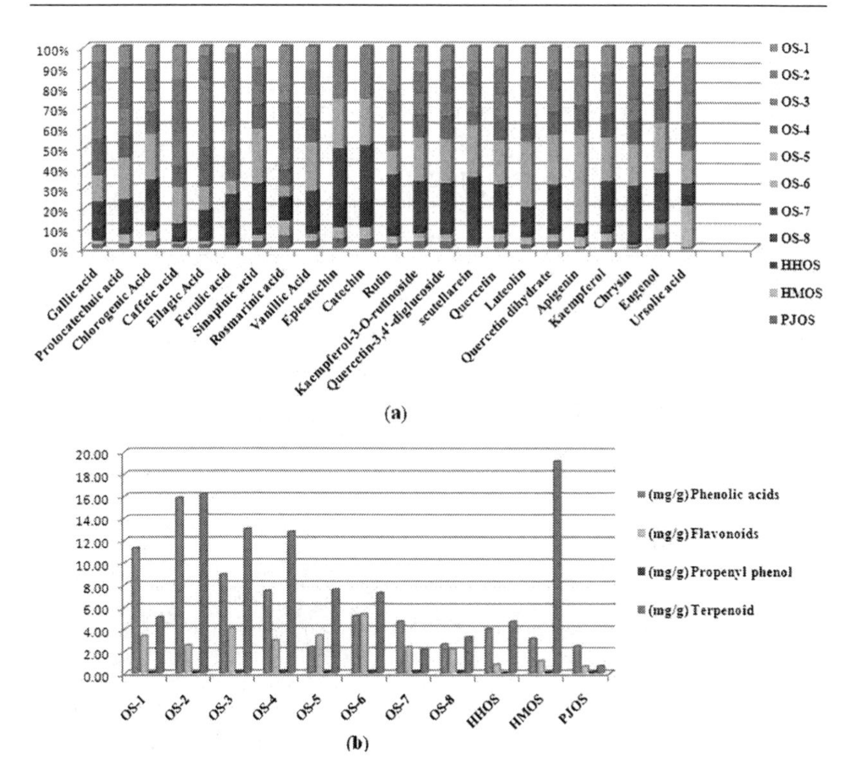

FIGURE 3.2 (a) Distribution of 23 bioactive markers in leaf extracts and marketed herbal formulations of *O. sanctum*. (b) Total content (mg/g) of phenolic acids, flavonoids, propenyl phenol and terpenoid in leaf extracts and marketed herbal formulations of *O. sanctum*. (Reproduced from Pandey et al. (2015) with permission from John Wiley and Sons.)

and catechin were found below the detection level in the leaf extracts from H. P. (OS-1), U. P. (OS-2, OS-4), M. P. (OS-6) and W. B. (OS-7) and scutellarein was found below the detection level in capsule (HHOS) and tablet (PJOS), whereas other bioactive markers were detected in all the analyzed samples. Ursolic acid (terpenoid) with a content range of 0.62–19.10 mg/g was the most abundant constituent in all the investigated samples except for tablet in which the most abundant component was rosmarinic acid (1.72 mg/g). Phenolic acids with a total content range of 2.33–15.78 mg/g were the second most abundant constituents in all the analyzed samples except for the leaf extract from M. P. (OS-5 and OS-6) in which flavonoids (3.39 and 5.33 mg/g, respectively) were the second most abundant constituents.

The total content of the nine phenolic acids was found to be the highest (15.78 mg/g) in the leaf extract from U. P. (OS-2) and lowest (2.33 mg/g) in the leaf extract from M. P. (OS-5). Similarly, the total content of 12 flavonoids was found to be the highest (5.33 mg/g) in the leaf extract from M. P. (OS-6) and lowest (0.60 mg/g) in the tablet (PJOS). The content of terpenoid (ursolic acid) was found to be the highest (19.10 mg/g) in the Himalayan *O. sanctum* (HMOS) capsule and lowest (0.62 mg/g) in the Patanjali *O. sanctum* (PJOS) tablet. Similarly, the content of propenyl phenol (eugenol) was found to be the highest (0.19 mg/g) in leaf extract from U. P. (OS-4) and lowest (0.02 mg/g) in the Herbal Hill *O. sanctum* (HHOS) capsule. The overall quantitative analysis results indicated that the leaf extract of *O. sanctum* from U. P. (OS-2) has the highest total content (34.44 mg/g) of 23 bioactive markers quantified in this study compared with other samples. The comprehensive quantitative analysis identified the variations in the chemical compositions of analyzed samples, and it might result in different degree of pharmacological activities.

3.10 DISCRIMINATION OF *O. SANCTUM* FROM DIFFERENT GEOGRAPHICAL REGIONS BY PCA

In this study, PCA was applied to statistically establish the correlation and discrimination of eight collections of *O. sanctum* representing different geographical regions of India, and to explore markers contributing to the classification. Although the chemical constituents of samples from different geographical regions were similar, the content of each compound varied significantly. The discrimination of these samples by visual inspection of data could be extremely tough and time consuming, and in contrast, by using a pattern recognition approach such as PCA, more subtle changes could be detected, and the analytical results are more reasonable.

The content of twenty three compounds in eight samples was determined simultaneously, and three replicates of each sample were used for PCA. The data matrix used was 24×23, where eight samples of *O. sanctum* replicated thrice for twenty three peaks/compounds. These twenty three peaks/compounds were used as variables for analysis. PCA was performed using correlation matrix of scaled data. Scaling of data matrix was carried out to adjust the mean and variance of each mass/peak so that the observations could become scale free and brought to a common platform for PCA. This pretreatment provides a mean value of 0 and variance of 1 to each mass. Scaling of data matrix such that all variables have zero mean and unit variance is also known as normalization. Score plot and the loading plot are displayed in Figure 3.3.

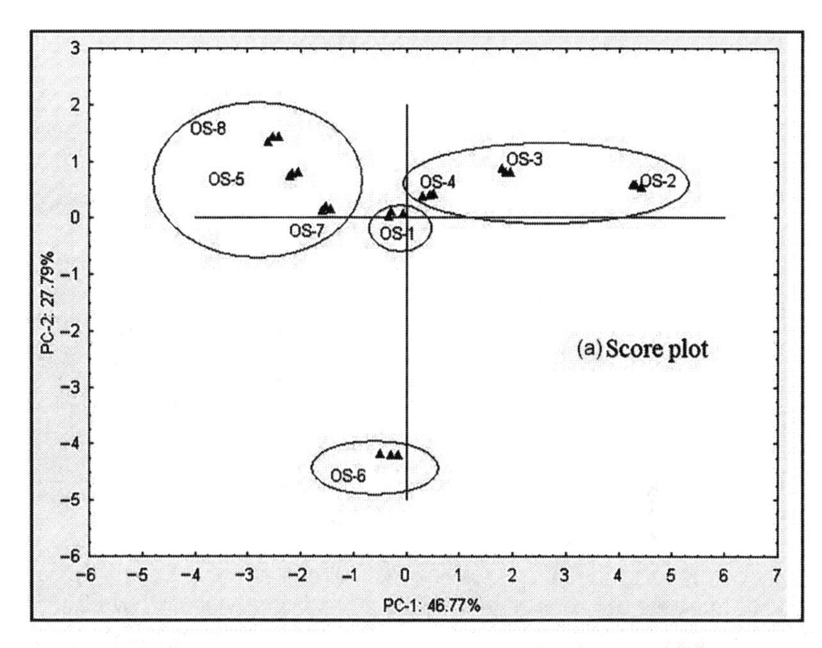

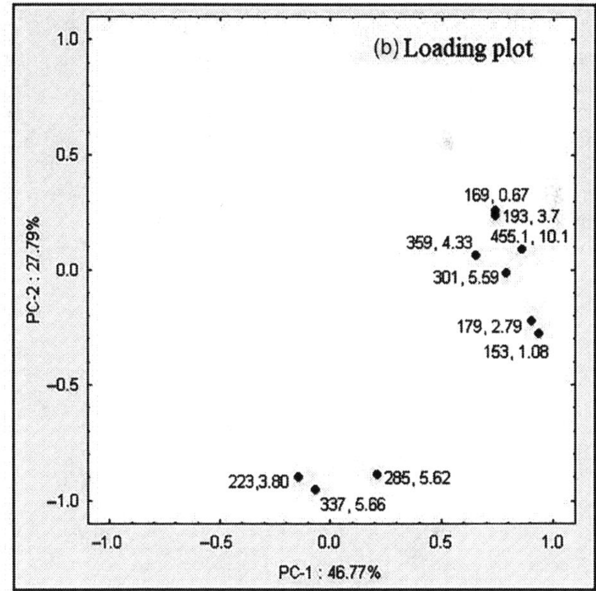

FIGURE 3.3 PCA (a) score plot and (b) loading plot. (Reproduced from Pandey et al. (2015) with permission from John Wiley and Sons.)

Ten out of twenty three analytes were able to be clearly identified in the eight samples with the dimensions reduced to three components.

The PC1 and PC2 together explained about 75% variation in the eight collections of *O. sanctum*. Score plot (Figure 3.3) showed the unique characteristics of three collections from U. P. (OS-2, OS-3 and OS-4). They are placed in the first quadrant with both PC1 and PC2 scores being positive. The sample from H. P. (OS-1) scores zero on PC2, and a negative score close to the origin on PC1 is its characteristic. The two samples from M. P. showed the highest variation in the quantities of compounds in the multidimensional space. The quantitative pattern of compounds in the sample from M. P. (OS-5) collected in 2011 was similar to W. B. However, the sample from M. P. (OS-6) collected in 2009 has a different quantitative pattern of compounds than its 2011 sample values. The distance between them indicates a variation in the quantities of compounds in these two samples, and this variation between years may be indicative of an overall different balance in the medicinal properties of the plant under changing climatic conditions. The two samples from W. B. (OS-7 and OS-8) fall in the second quadrant. The classification produced suggests different internal quality of samples from different sources. In the loading plot, the markers responsible for the cluster formation were mainly compounds (*m/z* 193, t_R 3.7 min; *m/z* 359, t_R 4.33 min; *m/z* 285, t_R 5.62 min; *m/z* 455.1, t_R 10.1 min) that suggested that the content of ferulic acid, rosmarinic acid, luteolin and ursolic acid had a significant relationship with sample sources.

Conclusion

4

This book covers phytochemical investigations of *Ocimum* spp. plants. It depicts method developments, and validation of simple, rapid, sensitive and accurate analytical methods for comparative screening and determination of phenolics and triterpenic acids in leaf extracts of six *Ocimum* species using HPLC-ESI-QTOF-MS/MS and UHPLC-QqQ_{LIT}-MS/MS. Fifty compounds were identified and characterized by their retention behavior, exact mass measurement, molecular formula, MS/MS spectral patterns and authentic standards. The developed UHPLC-QqQ_{LIT}-MS/MS method was effectively applied in leaf extracts of six *Ocimum* species to examine the variations in the content of 15 bioactive constituents. Quantitative outcomes demonstrated that ursolic acid and rosmarinic acid were the significant constituents of almost all the studied *Ocimum* species except for *O. americanum*. It was also observed that *O. killimandscharicum* contained the maximum amount of ursolic acid and eugenol and showed the highest total content of 15 bioactive phytochemicals compared to other samples. The comparative analysis of phenolics and triterpenic acids in leaf extracts of six *Ocimum* species may help consumers to select one of them as per their prerequisite.

Similarly, a UHPLC-ESI-MS/MS method was developed to study phytochemicals to standardize the traditionally important herbal medicines prepared using *O. sanctum*. The developed method was successfully applied to study the quantitative variation of twenty three bioactive markers in eight collections of *O. sanctum* collected from different geographical regions of India. High contents of terpenoids and phenolic acids were found in almost all the investigated samples. The PCA results revealed significant differences on the basis of quantitative analysis of selected phytochemicals among samples collected from different geographical regions of India. All the results obtained from this investigation demonstrated that the developed methods are simple, rapid, sensitive, accurate and precise for the screening and quantification of multiple bioactive constituents in *Ocimum* species, which offers a well acceptable methodology for authentication and quality control of *Ocimum* species and derived herbal formulations.

References

Adeniyi, S. A., C. L. Orjiekwe, J. E. Ehiagbonare, and B. D. Arimah. "Preliminary phytochemical analysis and insecticidal activity of ethanolic extracts of four tropical plants (*Vernonia amygdalina, Sida acuta, Ocimum gratissimum* and *Telfaria occidentalis*) against beans weevil (*Acanthscelides obtectus*)." *International Journal of Physical Sciences* 5, no. 6 (2010): 753–762.

Adigüzel, Ahmet, Medine Güllüce, Meryem Şengül, Hatice Öğütcü, Fikrettin Şahin, and İsa Karaman. "Antimicrobial effects of *Ocimum basilicum* (Labiatae) extract." *Turkish Journal of Biology* 29, no. 3 (2005): 155–160.

Akgül, A. "Volatile oil composition of sweet basil (*Ocimum basilicum* L.) cultivating in Turkey." *Food/Nahrung* 33, no. 1 (1989): 87–88.

Akinmoladun, Afolabi C., E. O. Ibukun, Emmanuel Afor, Efere Martins Obuotor, and E. O. Farombi. "Phytochemical constituent and antioxidant activity of extract from the leaves of *Ocimum gratissimum*." *Scientific Research and Essays* 2, no. 5 (2007): 163–166.

Alam, P., J. Gupta, S. Firdouse, A. Firdouse, and J. Afshan. "HPTLC method for qualitative and quantitative estimation of eugenol from *Ocimum sanctum* linn in polyherbal formulation." *International Journal of Pharmaceutics* 7 (2012): 1–3.

Ali, Huma, and Savita Dixit. "In vitro antimicrobial activity of flavanoids of *Ocimum sanctum* with synergistic effect of their combined form." *Asian Pacific Journal of Tropical Disease* 2 (2012): S396–S398.

Anand, Ankur Kumar, Manindra Mohan, S. Zafar Haider, and Akash Sharma. "Essential oil composition and antimicrobial activity of three Ocimum species from Uttarakhand (India)." *International Journal of Pharmacy and Pharmaceutical Sciences* 3, no. 3 (2011): 223–225.

Anandjiwala, Sheetal, Jyoti Kalola, and Mandapati Rajani. "Quantification of eugenol, luteolin, ursolic acid, and oleanolic acid in black (Krishna Tulasi) and green (Sri Tulasi) varieties of *Ocimum sanctum* Linn. using high-performance thin-layer chromatography." *Journal of AOAC International* 89, no. 6 (2006): 1467–1474.

Anonymous. "Medicinal Plants for Livestock, *Ocimum basilicum, O. americanum* and *O. micranthum*." Cornell University Animal Science, 2010. http://www.ansci.cornell.edu/plants/medicinal/basil.html.

Anonymous. "Ayurvedic pharmacopoeia of India." Government of India, Ministry of Ayush, Pharmacopoeia Commission for Indian Medicine & Homoeopathy Ghaziabad, New Delhi, 1, no. 9 (2016): 99–105.

Asha, M. K., D. Prashanth, B. Murali, R. Padmaja, and A. Amit. "Anthelmintic activity of essential oil of *Ocimum sanctum* and eugenol." *Fitoterapia* 72, no. 6 (2001): 669–670.

Atal, Chand Kumar, and B. M. Kapur (Editor). *Cultivation and Utilization of Medicinal Plants*. New Delhi: Regional Research Laboratory, Council of Scientific & Industrial Research, 1982.

Awasthi, P. K., and S. C. Dixit. "Chemical compositions of *Ocimum sanctum* Shyama and *Ocimum sanctum* Rama oils from the plains of Northern India." *Journal of Essential Oil Bearing Plants* 10, no. 4 (2007): 292–296.

Ba-Hamdan, Abeer Hamdan Abdullah, Aly Magda Mohammad, and Bafeel Sameera Omer. "Antimicrobial activities and phytochemical analysis of the essential oil of *Ocimum basilicum*, collected from Jeddah Region, Saudi Arabia." *Journal of Microbiology Research* 4, no. 6A (2014):1–9.

Barreto, Jacqueline C., Maria TS Trevisan, William E. Hull, Gerhard Erben, Edy S. De Brito, Beate Pfundstein, Gerd Würtele, Bertold Spiegelhalder, and Robert W. Owen. "Characterization and quantitation of polyphenolic compounds in bark, kernel, leaves, and peel of mango (Mangifera indica L.)." *Journal of Agricultural and Food Chemistry* 56, no. 14 (2008): 5599–5610.

Bassolé, Imaël Henri Nestor, Aline Lamien-Meda, Balé Bayala, Souleymane Tirogo, Chlodwig Franz, Johannes Novak, Roger Charles Nebié, and Mamoudou Hama Dicko. "Composition and antimicrobial activities of Lippia multiflora Moldenke, Mentha x piperita L. and *Ocimum basilicum* L. essential oils and their major monoterpene alcohols alone and in combination." *Molecules* 15, no. 11 (2010): 7825–7839.

Batta, S. K., and G. Santhakumari. "The antifertility effect of *Ocimum sanctum* and Hibiscus cosa sinensis." *Indian Journal of Medical Research* 59, no. 5 (1970): 777–781.

Beatovic, Damir, Dijana Krstic-Milosevic, Snezana Trifunovic, Jovana Siljegovic, Jasmina Glamoclija, Mihailo Ristic, and Slavica Jelacic. "Chemical composition, antioxidant and antimicrobial activities of the essential oils of twelve *Ocimum basilicum* L. cultivars grown in Serbia." *Records of Natural Products* 9, no. 1 (2015): 62.

Bhargava, K. P., and N. Singh. "Anti-stress activity of *Ocimum sanctum* Linn." *The Indian Journal of Medical Research* 73 (1981): 443.

Bhattacharya, S. K. *Handbook of Aromatic Plants*. Jaipur: Pointer Publications, 2004.

Bhatti, H. A. "Ph.D. Thesis, isolation and structure elucidation of the chemical constituents of *Ocimum basilicum* L. and Centella asiatica and synthesis of Lawsonin and its relative dhydrobenzofuran derivatives." International Center for Chemical and Biological Sciences H.E.J. Research Institue of Chemistry University of Karachi, Pakistan, pp. 1–217, 2008.

Bilal, Alia, Nasreen Jahan, Ajij Ahmed, Saima Naaz Bilal, Shahida Habib, and Syeda Hajra. "Phytochemical and pharmacological studies on *Ocimum basilicum* Linn-A review." *International Journal of Current Research and Review* 4, no. 23 (2012).

Budka, Dirk, and N. A. Khan. "The effect of *Ocimum basilicum*, Thymus vulgaris, Origanum vulgare essential oils on Bacillus cereus in rice-based foods." *EJBS* 2, no. 1 (2010): 17–20.

Bunrathep, Supawan, Chanida Palanuvej, and Nijsiri Ruangrungsi. "Chemical Compositions and Antioxidative Activities of Essential Oils from Four Ocimum Species Endemic to Thailand." *Journal of Health Research* 21, no. 3 (2007): 201–206.

Chalchat, Jean-Claude, Raymond-Philippe Garry, Lassine Sidibé, and Moussa Harama. "Aromatic plants of Mali (I): chemical composition of essential oils of *Ocimum basilicum* L." *Journal of Essential Oil Research* 11, no. 3 (1999): 375–380.

Chandra, Rashmi, Vinay Dwivedi, Kumar Shivam, and Abhimanyu Kumar Jha. "Detection of antimicrobial activity of Oscimum sanctum (Tulsi) & Trigonella foenum graecum (Methi) against some selected bacterial & fungal strains." *Research Journal of Pharmaceutical, Biological and Chemical Sciences* 2, no. 4 (2011): 809–813.

Chen, Shu-Yun, Ting-Xuan Dai, Yuan-Tsung Chang, Shyh-Shyan Wang, Shang-Ling Ou, Wen-Ling Chuang, Chih-Yun Chuang, Yi-Hua Lin, Yi-Yin Lin, and Hsin-Mei Ku. "Genetic diversity among '*Ocimum*' species based on ISSR, RAPD and SRAP markers." *Australian Journal of Crop Science* 7, no. 10 (2013): 1463.

Choudhury, G. B., M. Behera, S. K. Tripathy, P. K. Jena, and S. R. Mishra. "UV, HPLC Method development and quantification of eugenol isolated by preparative paper chromatography from Alcoholic extracts of different species of *Ocimum*." *International Journal of Chemical and Analytical Science* 2 (2011): 3–6.

Clayton, W. D. *Flora of Tropical East Africa. Gramineae (Part 1)*. London: Crown Agents for Oversea Governments and Administrations, 1970.

Cuyckens, F., Y. L. Ma, G. Pocsfalvi, and M. Claeysi. "Tandem mass spectral strategies for the structural characterization of flavonoid glycosides." *Analusis* 28, no. 10 (2000): 888–895.

Darrah, Helen H. "Investigation of the cultivars of the Basils (Ocimum)." *Economic Botany* 28, no. 1 (1974): 63–67.

de Almeida, Igor, Daniela Sales Alviano, Danielle Pereira Vieira, Péricles Barreto Alves, Arie Fitzgerald Blank, Angela Hampshire CS Lopes, Celuta Sales Alviano, and S. Rosa Maria do Socorro. "Antigiardial activity of *Ocimum basilicum* essential oil." *Parasitology Research* 101, no. 2 (2007): 443–452.

De Martino, Laura, Vincenzo De Feo, and Filomena Nazzaro. "Chemical composition and in vitro antimicrobial and mutagenic activities of seven Lamiaceae essential oils." *Molecules* 14, no. 10 (2009): 4213–4230.

Deo, S. S., F. Inam, and R. P. Mahashabde. "Antimicrobial activity and HPLC fingerprinting of crude ocimum extracts." *Journal of Chemistry* 8, no. 3 (2011): 1430–1437.

Devi, P. Uma, K. S. Bisht, and M. Vinitha. "A comparative study of radioprotection by Ocimum flavonoids and synthetic aminothiol protectors in the mouse." *The British Journal of Radiology* 71, no. 847 (1998): 782–784.

Dey, B. B., and M. A. Choudhuri. "Effect of Leaf Development Stage on Changes in Essential Oil of *Ocimum sanctum* L." *Biochemie und Physiologie der Pflanzen* 178, no. 5 (1983): 331–335.

Edris, Amr E., and Eman S. Farrag. "Antifungal activity of peppermint and sweet basil essential oils and their major aroma constituents on some plant pathogenic fungi from the vapor phase." *Food/Nahrung* 47, no. 2 (2003): 117–121.

Erler, F., I. Ulug, and B. Yalcinkaya. "Repellent activity of five essential oils against Culex pipiens." *Fitoterapia* 77, no. 7–8 (2006): 491–494.

Eshraghian, Ahad. "Anti-Inflammatory, gastrointestinal and hepatoprotective effects of *Ocimum sanctum* Linn: an ancient remedy with new application." *Inflammation & Allergy-Drug Targets (Formerly Current Drug Targets-Inflammation & Allergy)* 12, no. 6 (2013): 378–384.

Fabre, Nicolas, Isabelle Rustan, Edmond de Hoffmann, and Joëlle Quetin-Leclercq. "Determination of flavone, flavonol, and flavanone aglycones by negative ion liquid chromatography electrospray ion trap mass spectrometry." *Journal of the American Society for Mass Spectrometry* 12, no. 6 (2001): 707–715.

Fang, Nianbai, Shanggong Yu, and Ronald L. Prior. "LC/MS/MS characterization of phenolic constituents in dried plums." *Journal of Agricultural and Food Chemistry* 50, no. 12 (2002): 3579–3585.

Farivar, Taghi Naserpour. "In vitro and macrophage culture." *Journal of Medical Sciences* 6, no. 3 (2006): 348–351.

Fathiazad, Fatemeh, Amin Matlobi, Arash Khorrami, Sanaz Hamedeyazdan, Hamid Soraya, Mojtaba Hammami, Nasrin Maleki-Dizaji, and Alireza Garjani. "Phytochemical screening and evaluation of cardioprotective activity of ethanolic extract of *Ocimum basilicum* L.(basil) against isoproterenol induced myocardial infarction in rats." *DARU Journal of Pharmaceutical Sciences* 20, no. 1 (2012): 87.

Grayer, Renée J., Geoffrey C. Kite, Mamdouh Abou-Zaid, and Louise J. Archer. "The application of atmospheric pressure chemical ionisation liquid chromatography–mass spectrometry in the chemotaxonomic study of flavonoids: characterisation of flavonoids from *Ocimum gratissimum* var. gratissimum." *Phytochemical Analysis: An International Journal of Plant Chemical and Biochemical Techniques* 11, no. 4 (2000): 257–267.

Grayer, Renée J., Geoffrey C. Kite, Nigel C. Veitch, Maria R. Eckert, Petar D. Marin, Priyanganie Senanayake, and Alan J. Paton. "Leaf flavonoid glycosides as chemosystematic characters in *Ocimum*." *Biochemical Systematics and Ecology* 30, no. 4 (2002): 327–342.

Grayer, Renée J., Nigel C. Veitch, Geoffrey C. Kite, Anna M. Price, and Tetsuo Kokubun. "Distribution of 8-oxygenated leaf-surface flavones in the genus *Ocimum*." *Phytochemistry* 56, no. 6 (2001): 559–567.

Gu, Dongyu, Yi Yang, Mahinur Bakri, Qibin Chen, Xuelei Xin, and Haji Akber Aisa. "A LC/QTOF–MS/MS application to investigate chemical compositions in a fraction with protein tyrosine phosphatase 1B inhibitory activity from Rosa rugosa flowers." *Phytochemical Analysis* 24, no. 6 (2013): 661–670.

Gupta, Arti, Navin R. Sheth, Sonia Pandey, Dinesh R. Shah, and Jitendra S. Yadav. "Determination of ursolic acid in fractionated leaf extracts of *Ocimum gratissimum* Linn and in developed herbal hepatoprotective tablet by HPTLC." *Pharmacognosy Journal* 5, no. 4 (2013): 156–162.

Gupta, Prasoon, Dinesh Kumar Yadav, Kiran Babu Siripurapu, Guatam Palit, and Rakesh Maurya. "Constituents of *Ocimum sanctum* with antistress activity." *Journal of Natural Products* 70, no. 9 (2007): 1410–1416.

Gupta, S. K., Jai Prakash, and Sushma Srivastava. "Validation of traditional claim of Tulsi, *Ocimum sanctum* Linn. as a medicinal plant." *Indian Journal of Experimental Biology* 40, no. 7 (2002):765–773.

Hakkim, F. Lukmanul, C. Gowri Shankar, and S. Girija. "Chemical composition and antioxidant property of holy basil (*Ocimum sanctum* L.) leaves, stems, and inflorescence and their in vitro callus cultures." *Journal of Agricultural and Food Chemistry* 55, no. 22 (2007): 9109–9117.

Hossain, M. Amzad, M. J. Kabir, S. M. Salehuddin, SM Mizanur Rahman, A. K. Das, Sandip Kumar Singha, Md Khorshed Alam, and Atiqur Rahman. "Antibacterial properties of essential oils and methanol extracts of sweet basil *Ocimum basilicum* occurring in Bangladesh." *Pharmaceutical Biology* 48, no. 5 (2010a): 504–511.

Hossain, Mohammad B., Dilip K. Rai, Nigel P. Brunton, Ana B. Martin-Diana, and Catherine Barry-Ryan. "Characterization of phenolic composition in Lamiaceae spices by LC-ESI-MS/MS." *Journal of Agricultural and Food Chemistry* 58, no. 19 (2010b): 10576–10581.

ICH Harmonised Tripartite. "Validation of analytical procedures: text and methodology Q2 (R1) guidelines." *International Conference on Harmonization*, Geneva, Switzerland, 2005, pp. 11–12.

Ismail, M. "Central properties and chemical composition of *Ocimum basilicum* essential oil." *Pharmaceutical Biology* 44, no. 8 (2006): 619–626.

Itankar, Prakash R., Mohammad Tauqeer, Jayshree S. Dalal, and Pranali G. Chatole. "Simultaneous determination of ursolic acid and eugenol from *Ocimum sanctum* L. cultivated by organic and non-organic farming." *Indian Journal of Traditional Knowledge* 14, no. 4 (2015): 620–625.

Joshi, R. K. "In vitro antimicrobial and antioxidant activities of the essential oils of *Ocimum gratissimum*, O. sanctum and their major constituents." *Indian Journal of Pharmaceutical Sciences* 75, no. 4 (2013): 457–462.

Joshi, Rakesh Kumar. "Phytoconstituents, traditional, medicinal and bioactive uses of Tulsi (*Ocimum sanctum* Linn.): a review." *Journal of Pharmacognosy and Phytochemistry* 6, no. 2 (2017): 261–264.

Kasali, Adeleke A., Adeolu O. Eshilokun, Segun Adeola, Peter Winterhalter, Holger Knapp, Bernd Bonnlander, and Wilfried A. Koenig. "Volatile oil composition of new chemotype of *Ocimum basilicum* L. from Nigeria." *Flavour and Fragrance Journal* 20, no. 1 (2005): 45–47.

Kaya, Ilhan, Nazife Yigit, and Mehlika Benli. "Antimicrobial activity of various extracts of *Ocimum basilicum* L. and observation of the inhibition effect on bacterial cells by use of scanning electron microscopy." *African Journal of Traditional, Complementary and Alternative Medicines* 5, no. 4 (2008): 363–369.

Kéita, Sékou Moussa, Charles Vincent, Jean-Pierre Schmit, and André Bélanger. "Essential oil composition of *Ocimum basilicum* L., *O. gratissimum* L. and *O. suave* L. in the Republic of Guinea." *Flavour and Fragrance Journal* 15, no. 5 (2000): 339–341.

Kelm, M. A., M. G. Nair, G. M. Strasburg, and D. L. DeWitt. "Antioxidant and cyclooxygenase inhibitory phenolic compounds from *Ocimum sanctum* Linn." *Phytomedicine* 7, no. 1 (2000): 7–13.

Kelm, Mark A., and Muraleedharan G. Nair. "Mosquitocidal compounds and a triglyceride, 1, 3-dilinoleneoyl-2-palmitin, from *Ocimum sanctum*." *Journal of Agricultural and Food Chemistry* 46, no. 8 (1998): 3092–3094.

Khanna, N., and Jagriti Bhatia. "Antinociceptive action of *Ocimum sanctum* (Tulsi) in mice: possible mechanisms involved." *Journal of Ethnopharmacology* 88, no. 2–3 (2003): 293–296.

Khatri, L. M., M. K. A. Nasir, R. Saleem, and F. Noor. "Evaluation of Pakistani sweet basil oil for commercial exploitation." *Pakistan Journal of Scientific and Industrial Research* 38 (1995): 281–282.

Kitchlu, S., Rekha Bhadauria, Gandhi Ram, Kushal Bindu, Ravi K. Khajuria, and Ashok Ahuja. "Chemo-divergence in essential oil composition among thirty one core collections of *Ocimum sanctum* L. grown under sub-tropical region of Jammu, India." *American Journal of Plant Sciences* 4, no. 02 (2013): 302.

Koba, Koffi, P. W. Poutouli, Christine Raynaud, Jean-Pierre Chaumont, and Komla Sanda. "Chemical composition and antimicrobial properties of different basil essential oils chemotypes from Togo." *Bangladesh Journal of Pharmacology* 4, no. 1 (2009): 1–8.

Kothari, S. K., A. K. Bhattacharya, and S. Ramesh. "Essential oil yield and quality of methyl eugenol rich Ocimum tenuiflorum Lf (syn. O. sanctum L.) grown in south India as influenced by method of harvest." *Journal of Chromatography A* 1054, no. 1–2 (2004): 67–72.

Kristinsson, Karl G., Anna B. Magnusdottir, Hannes Petersen, and Ann Hermansson. "Effective treatment of experimental acute otitis media by application of volatile fluids into the ear canal." *The Journal of Infectious Diseases* 191, no. 11 (2005): 1876–1880.

Kumar, Anant, Karishma Agarwal, Anil Kumar Maurya, Karuna Shanker, Umme Bushra, Sudeep Tandon, and Dnyaneshwar U. Bawankule. "Pharmacological and phytochemical evaluation of *Ocimum sanctum* root extracts for its antiinflammatory, analgesic and antipyretic activities." *Pharmacognosy Magazine* 11, no. Suppl. 1 (2015): S217.

Kumar, P. K., M. R. Kumar, K. Kavitha, J. Singh, and R. Khan. "Pharmacological actions of *Ocimum sanctum*–review article." *International Journal of Advances in Pharmacy, Biology and Chemistry* 1, no. 3 (2012): 2277–4688.

Lalla, Jogender, Purnima Hamrapurkar, and Abhishek Singh. "Quantitative HPTLC analysis of the eugenol content of leaf powder and a capsule formulation of *Ocimum sanctum.*" *JPC-Journal of Planar Chromatography-Modern TLC* 20, no. 2 (2007): 135–138.

Laskar, S., and S. Ghosh Majumdar. "Variation of major constituents of essential oil of the leaves of ocimum-sanctum linn." *Journal of the Indian Chemical Society* 65, no. 4 (1988): 301–302.

Ling Chang, Chiou, Il Kyu Cho, and Qing X. Li. "Insecticidal activity of basil oil, trans-anethole, estragole, and linalool to adult fruit flies of Ceratitis capitata, Bactrocera dorsalis, and Bactrocera cucurbitae." *Journal of Economic Entomology* 102, no. 1 (2009): 203–209.

Mahajan, Nipun, Shruti Rawal, Monika Verma, Mayur Poddar, and Shashi Alok. "A phytopharmacological overview on Ocimum species with special emphasis on *Ocimum sanctum.*" *Biomedicine & Preventive Nutrition* 3, no. 2 (2013): 185–192.

Mandal, Shyamapada, Manisha Deb Mandal, and Nishith Kumar Pal. "Enhancing chloramphenicol and trimethoprim in vitro activity by *Ocimum sanctum* Linn.(Lamiaceae) leaf extract against Salmonella enterica serovar Typhi." *Asian Pacific Journal of Tropical Medicine* 5, no. 3 (2012): 220–224.

Marotti, Mauro, Roberta Piccaglia, and Enrico Giovanelli. "Differences in essential oil composition of basil (*Ocimum basilicum* L.) Italian cultivars related to morphological characteristics." *Journal of Agricultural and Food Chemistry* 44, no. 12 (1996): 3926–3929.

Mishra, M. "Tulsi to save Taj Mahal from pollution effects." The Times of India, Bennett Coleman and Co. Ltd., 2008.

Mishra, Poonam, and Sanjay Mishra. "Study of antibacterial activity of *Ocimum sanctum* extract against gram positive and gram negative bacteria." *American Journal of Food Technology* 6, no. 4 (2011): 336–341.

Mohammed, Mohiuddin, M. J. Chowdhury, M. K. Alam, and M. K. Hossain. "Chemical composition of essential oil of four flavouring plants used by the tribal people of Bandarban hill district in Bangladesh." *International Journal of Medicinal and Aromatic Plants* 2, no. 1 (2012): 106–113.

Mohan, Lalit, M. V. Amberkar, and Meena Kumari. "*Ocimum sanctum* Linn (Tulsi)—an overview." *International Journal of Pharmaceutical Sciences Review and Research* 7, no. 1 (2011): 51–53.

Mondal, Shankar, Bijay R. Mirdha, and Sushil C. Mahapatra. "The science behind sacredness of Tulsi (*Ocimum sanctum* Linn.)." *Indian Journal of Physiology* and *Pharmacology* 53, no. 4 (2009): 291–306.

Mondello, Luigi, Giovanni Zappia, Antonella Cotroneo, Ivana Bonaccorsi, Jasim Uddin Chowdhury, Mohammed Yusuf, and Giovanni Dugo. "Studies on the essential oil-bearing plants of Bangladesh. Part VIII. Composition of some *Ocimum* oils *O. basilicum* L. var. purpurascens; *O. sanctum* L. green; *O. sanctum* L. purple; *O. americanum* L., citral type; *O. americanum* L., camphor type." *Flavour and Fragrance Journal* 17, no. 5 (2002): 335–340.

Nagarjun, S., H. C. Jain, and G. C. Aulakh. "Indigenous plant used in fertility control, in cultivation & utilization of medicinal plants." In C. K. Atal and B. M. Kapoor (eds.), *Cultivation and Utilization of Medicinal Plants* (p. 558). New Delhi: PID CSIR, 1989.

Nakamura, Celso Vataru, Kelly Ishida, Ligia Carla Faccin, Benedito Prado Dias Filho, Diógenes Aparício Garcia Cortez, Sonia Rozental, Wanderley de Souza, and Tânia Ueda-Nakamura. "In vitro activity of essential oil from *Ocimum gratissimum* L. against four Candida species." *Research in Microbiology* 155, no. 7 (2004): 579–586.

Nakamura, Celso Vataru, Tania Ueda-Nakamura, Erika Bando, Abrahão Fernandes Negrão Melo, Díogenes Aparício Garcia Cortez, and Benedito Prado Dias Filho. "Antibacterial activity of *Ocimum gratissimum* L. essential oil." *Memórias do Instituto Oswaldo Cruz* 94, no. 5 (1999): 675–678.

Nguefack, J., Birgitte Bjørn Budde, and Mogens Jakobsen. "Five essential oils from aromatic plants of Cameroon: their antibacterial activity and ability to permeabilize the cytoplasmic membrane of Listeria innocua examined by flow cytometry." *Letters in Applied Microbiology* 39, no. 5 (2004): 395–400.

Ntezurubanza, L., J. J. C. Scheffer, and A. Baerheim Svendsen. "Composition of the essential oil of *Ocimum gratissimum* grown in rwandal." *Planta Medica* 53, no. 5 (1987): 421–423.

Okigbo, R. N., C. S. Mbajiuka, and C. O. Njoku. "Antimicrobial potential of (UDA) Xylopia aethopica and *Ocimum gratissimum* L. on some pathogens of man." *International Journal of Molecular Medicine and Advance Sciences* Pakistan 1, no. 4 (2005): 392–394.

Onwuliri, F. C., D. L. Wonang, and E. A. Onwuliri. "Phytochemical, toxicological and histo-pathological studies of some medicinal plants in Nigeria." *IRJNS* 2, no. (2006): 225–229.

Opalchenova, G., and D. Obreshkova. "Modern phytomedicine: turning medicinal plants into drugs." *Journal of Microbiological Methods* 54, no. (2003): 105–110.

Ovesná, Zdenka, Katarína Kozics, and Darina Slameňová. "Protective effects of ursolic acid and oleanolic acid in leukemic cells." *Mutation Research/Fundamental and Molecular Mechanisms of Mutagenesis* 600, no. 1–2 (2006): 131–137.

Özcan, Musa, and Jean-Clause Chalchat. "Essential oil composition of *Ocimum basilicum* L." *Czech Journal of Food Sciences* 20, no. 6 (2002): 223–228.

Pandey, Abhay Kumar, Pooja Singh, and Nijendra Nath Tripathi. "Chemistry and bioactivities of essential oils of some Ocimum species: an overview." *Asian Pacific Journal of Tropical Biomedicine* 4, no. 9 (2014): 682–694.

Pandey, B. P., and Anita. *Economic Botany* (p. 294). New Delhi: Chand & Company Ltd., 1990.

Pandey, Renu, and Brijesh Kumar. "HPLC–QTOF–MS/MS-based rapid screening of phenolics and triterpenic acids in leaf extracts of Ocimum species and their interspecies variation." *Journal of Liquid Chromatography & Related Technologies* 39, no. 4 (2016): 225–238.

Pandey, Renu, Preeti Chandra, Brijesh Kumar, Bhupender Dutt, and Kulwant Rai Sharma. "A rapid and highly sensitive method for simultaneous determination of bioactive constituents in leaf extracts of six Ocimum species using ultra high performance liquid chromatography-hybrid linear ion trap triple quadrupole mass spectrometry." *Analytical Methods* 8, no. 2 (2016): 333–341.

Pandey, Renu, Preeti Chandra, Mukesh Srivastava, D. K. Mishra, and Brijesh Kumar. "Simultaneous quantitative determination of multiple bioactive markers in *Ocimum sanctum* obtained from different locations and its marketed herbal formulations using UPLC-ESI-MS/MS combined with principal component analysis." *Phytochemical Analysis* 26, no. 6 (2015): 383–394.

Parag, Sadgir, Nilosey Vijyayshree, Bhandari Ranu, and B. R. Patil. "Antibacterial activity of *Ocimum sanctum* Linn and its application in water purification." *Research Journal of Chemistry and Environment* 14, no. 3 (2010): 46–50.

Parasuraman, Subramani, Subramani Balamurugan, Parayil Varghese Christapher, Rajendran Ramesh Petchi, Wong Yeng Yeng, Jeyabalan Sujithra, and Chockalingam Vijaya. "Evaluation of antidiabetic and antihyperlipidemic effects of hydroalcoholic extracts of leaves of Ocimum tenuiflorum (Lamiaceae) and prediction of biological activity of its phytoconstituents." *Pharmacognosy Research* 7, no. 2 (2015): 156.

Patil, Raju, Ravindra Patil, Bharati Ahirwar, and Dheeraj Ahirwar. "Isolation and characterization of anti-diabetic component (bioactivity—guided fractionation) from *Ocimum sanctum* L.(Lamiaceae) aerial part." *Asian Pacific Journal of Tropical Medicine* 4, no. 4 (2011): 278–282.

Pattanayak, Priyabrata, Pritishova Behera, Debajyoti Das, and Sangram K. Panda. "*Ocimum sanctum* Linn. A reservoir plant for therapeutic applications: An overview." *Pharmacognosy Reviews* 4, no. 7 (2010): 95–105.

Patwardhan, Bhushan, Dnyaneshwar Warude, Palpu Pushpangadan, and Narendra Bhatt. "Ayurveda and traditional Chinese medicine: a comparative overview." *Evidence-Based Complementary and Alternative Medicine* 2, no. 4 (2005): 465–473.

Pavela, R., and T. Chermenskaya. "Potential insecticidal activity of extracts from 18 species of medicinal plants on larvae of Spodoptera littoralis." *Plant Protection Science* 40, no. 4 (2004): 145–150.

Pessoa, L. M., S. M. Morais, C. M. L. Bevilaqua, and J. H. S. Luciano. "Anthelmintic activity of essential oil of *Ocimum gratissimum* Linn. and eugenol against Haemonchus contortus." Veterinary parasitology 109, no. 1–2 (2002): 59–63.

Politeo, Olivera, M. Jukic, and M. Milos. "Chemical composition and antioxidant capacity of free volatile aglycones from basil (*Ocimum basilicum* L.) compared with its essential oil." Food Chemistry 101, no. 1 (2007): 379–385.

Prajapati, N. D., S. S. Purohit, A. K. Sharma, and T. A. Kumar. *Hand Book of Medicinal Plant* (p. 367). Jodhpur: Agrobios, 2003.

Prasad, M. P., K. Jayalakshmi, and G. G. Rindhe. "Antibacterial activity of Ocimum species and their phytochemical and antioxidant potential." *International Journal of Microbiology Research* 4, no. 8 (2012): 302.

Ramesh, B., and V. N. Satakopan. "In vitro Antioxidant Activities of Ocimum Species: *Ocimum basilicum* and *Ocimum sanctum*." *Journal of Cell & Tissue Research* 10, no. 1 (2010): 1–5.

Reuveni, A., R. Fleischer, and E. Putievsky. "Fungistatic activity of essential oils from *Ocimum basilicum* chemotypes." *Journal of Phytopathology* 110, no. 1 (1984): 20–22.

Rout, Kedar Kumar, Rajesh Kumar Singh, Durga Prasad Barik, and Sagar Kumar Mishra. "Thin-Layer Chromatographie Separation and Validated HPTLC Method for Quantification of Ursolic Acid in Various Ocimum Species." *Journal of Food & Drug Analysis* 20, no. 4 (2012): 865–871.

Runyoro, D., O. Ngassapa, K. Vagionas, N. Aligiannis, K. Graikou, and I. Chinou. "Chemical composition and antimicrobial activity of the essential oils of four *Ocimum* species growing in Tanzania." *Food Chemistry* 119, no. 1 (2010): 311–316.

Saini, Archana, Saroj Sharma, and Sanjay Chhibber. "Induction of resistance to respiratory tract infection with Klebsiella pneumoniae in mice fed on a diet supplemented with tulsi (*Ocimum sanctum*) and clove (Syzgium aromaticum) oils." *Journal of Microbiology, Immunology and Infection* 42, no. 2 (2009): 107–113.

Sajjadi, Seyed Ebrahim. "Analysis of the essential oils of two cultivated basil (*Ocimum basilicum* L) from Iran." *DARU Journal of Pharmaceutical Sciences* 14, no. 3 (2006): 128–130.

Samant, Sher Singh, Uppeandra Dhar, and Lok Man S. Palni. *Medicinal Plants of Indian Himalaya.* Kanpur: Gyanodaya Prakashan, 1998.

Santoro, Giani F., Maria G. Cardoso, Luiz Gustavo L. Guimarães, Lidiany Z. Mendonça, and Maurilio J. Soares. "Trypanosoma cruzi: activity of essential oils from Achillea millefolium L., Syzygium aromaticum L. and *Ocimum basilicum* L. on epimastigotes and trypomastigotes." *Experimental Parasitology* 116, no. 3 (2007): 283–290.

Sarkar, Depanjan, Amitava Srimany, and T. Pradeep. "Rapid identification of molecular changes in tulsi (*Ocimum sanctum* Linn) upon ageing using leaf spray ionization mass spectrometry." *Analyst* 137, no. 19 (2012): 4559–4563.

Sastry, Kakaraparthi Pandu, Ramachandran Ramesh Kumar, Arigari Niranjan Kumar, Gogte Sneha, and Margaret Elizabeth. "Morpho-chemical description and antimicrobial activity of different Ocimum species." *Journal of Plant Development* 19 (2012): 53–64.

Sawarkar, H. A., S. S. Khadabadi, D. M. Mankar, I. A. Farooqui, and N. S. Jagtap. "Development and biological evaluation of herbal anti-acne gel." *International Journal of PharmTech Research* 2, no. 3 (2010): 2028–2031.

Schulz, H., B. Schrader, R. Quilitzsch, S. Pfeffer, and H. Krüger. "Rapid classification of basil chemotypes by various vibrational spectroscopy methods." *Journal of Agricultural and Food Chemistry* 51, no. 9 (2003): 2475–2481.

Sen, P. "Therapeutic potentials of Tulsi: from experience to facts." *Drugs News & Views* 1, no. 2 (1993): 15–21.

Shokeen, Poonam, Krishna Ray, Manju Bala, and Vibha Tandon. "Preliminary studies on activity of *Ocimum sanctum*, Drynaria quercifolia, and Annona squamosa against Neisseria gonorrhoeae." *Sexually Transmitted Diseases* 32, no. 2 (2005): 106–111.

Shukla, S. T., V. H. Kulkarni, P. V. Habbu, K. S. Jagadeesh, B. S. Patil, and D. M. Smita. "Hepatoprotective and antioxidant activities of crude fractions of endophytic fungi of *Ocimum sanctum* Linn. in rats." *Oriental Pharmacy and Experimental Medicine* 12, no. 2 (2012): 81–91.

Siddiqui, H. H. "Safety of herbal drugs-an overview." *Drugs News & Views* 1, no. 2 (1993): 7–10.

Silva, M. Goretti, Icaro Vieira, Francisca Mendes, Irineu Albuquerque, Rogério Dos Santos, Fábio Silva, and Selene Morais. "Variation of ursolic acid content in eight Ocimum species from northeastern Brazil." *Molecules* 13, no. 10 (2008): 2482–2487.

Singh, D. K., and P. K. Hajra. "Floristic diversity." In G. S. Gujral and V. Sharma (eds.), *Changing Perspective of Biodiversity Status in the Himalaya* (pp. 23–38). New Delhi: British Council Division, British High Commission Publication, Wildlife Youth Services, 1996.

Srivastava, H. C., Ashok Srivastava, Pankaj Shukla, Ajay Singh Maurya, and Sonia Tripathi. "Analysis of the essential oils of two different cultivated basil (*Ocimum basilicum* L) from India." *International Journal of Pharmaceutical Sciences and Research* 4, no. 4 (2013): 1398.

Sun, Jianping, Feng Liang, Yan Bin, Ping Li, and Changqing Duan. "Screening non-colored phenolics in red wines using liquid chromatography/ultraviolet and mass spectrometry/mass spectrometry libraries." *Molecules* 12, no. 3 (2007): 679–693.

Sundaram, R. Shanmuga, M. Ramanathan, R. Rajesh, B. Satheesh, and Dhandayutham Saravanan. "LC-MS quantification of rosmarinic acid and ursolic acid in the *Ocimum sanctum* Linn. leaf extract (Holy basil, Tulsi)." *Journal of Liquid Chromatography & Related Technologies* 35, no. 5 (2012): 634–650.

Sundaram, Shanmuga. "Quantification of bioactive principles in Indian traditional herb *Ocimum sanctum* Linn.(Holy Basil) Leaves by high performance liquid chromatography." *Asian Journal of Biomedical and Pharmaceutical Sciences* 1, no. 3 (2011).

Suppakul, Panuwat, Joseph Miltz, Kees Sonneveld, and Stephen W. Bigger. "Active packaging technologies with an emphasis on antimicrobial packaging and its applications." *Journal of Food Science* 68, no. 2 (2003a): 408–420.

Suppakul, Panuwat, Joseph Miltz, Kees Sonneveld, and Stephen W. Bigger. "Antimicrobial properties of basil and its possible application in food packaging." *Journal of Agricultural and Food Chemistry* 51, no. 11 (2003b): 3197–3207.

Suzuki, Akiko, Osamu Shirota, Kanami Mori, Setsuko Sekita, Hiroyuki Fuchino, Akihito Takano, and Masanori Kuroyanagi. "Leishmanicidal active constituents from Nepalese medicinal plant Tulsi (*Ocimum sanctum* L.)." *Chemical and Pharmaceutical Bulletin* 57, no. 3 (2009): 245–251.

Tchoumbougnang, F., P. H. Amvam Zollo, E. Dagne, and Y. Mekonnen. "In vivo antimalarial activity of essential oils from Cymbopogon citratus and *Ocimum gratissimum* on mice infected with Plasmodium berghei." *Planta Medica* 71, no. 1 (2005): 20–23.

Thyagaraj, Vishruta Domlur, Rojison Koshy, Monica Kachroo, Anand S. Mayachari, Laxman P. Sawant, and Murali Balasubramanium. "A validated RP-HPLC-UV/DAD method for simultaneous quantitative determination of rosmarinic acid and eugenol in *Ocimum sanctum* L." *Pharmaceutical Methods* 4, no. 1 (2013): 1–5.

Trivedi, P. C. *Ethno Medical Plants of India*. Chicago, IL: University of Chicago, 2007.

Vani, S. Raseetha, S. F. Cheng, and C. H. Chuah. "Comparative study of volatile compounds from genus Ocimum." *American Journal of Applied Sciences* 6, no. 3 (2009): 523.

Verma, Ram Swaroop, Pawan Singh Bisht, Rajendra Chandra Padalia, Dharmendra Saikia, and Amit Chauhan. "Chemical composition and antibacterial activity of essential oil from two Ocimum spp. grown in sub-tropical India during spring-summer cropping season." *Journal of Traditional Medicines* 6, no. 5 (2011): 211–217.

Verma, Sunita. "Chemical constituents and pharmacological action of *Ocimum sanctum* (Indian holy basil-Tulsi)." *The Journal of Phytopharmacology* 5, no. 5 (2016): 205–207.

Vieira, Priscila RN, Selene M. de Morais, Francisco HQ Bezerra, Pablito Augusto Travassos Ferreira, Írvila R. Oliveira, and Maria Goretti V. Silva. "Chemical composition and antifungal activity of essential oils from Ocimum species." *Industrial Crops and Products* 55 (2014): 267–271.

Wu, Haifeng, Jian Guo, Shilin Chen, Xin Liu, Yan Zhou, Xiaopo Zhang, and Xudong Xu. "Recent developments in qualitative and quantitative analysis of phytochemical constituents and their metabolites using liquid chromatography–mass spectrometry." *Journal of Pharmaceutical and Biomedical Analysis* 72 (2013): 267–291.

Xia, Yuanyuan, Guangli Wei, Duanyun Si, and Changxiao Liu. "Quantitation of ursolic acid in human plasma by ultra performance liquid chromatography tandem mass spectrometry and its pharmacokinetic study." *Journal of Chromatography B* 879, no. 2 (2011): 219–224.

Ying, Xuhui, Mingying Liu, Qionglin Liang, Min Jiang, Yiming Wang, Fukai Huang, Yuanyuan Xie, Jie Shao, Gang Bai, and Guoan Luo. "Identification and analysis of absorbed components and their metabolites in rat plasma and tissues after oral administration of 'Ershiwuwei Shanhu'pill extracts by UPLC-DAD/Q-TOF-MS." *Journal of Ethnopharmacology* 150, no. 1 (2013): 324–338.

Yucharoen, Raenu, Songyot Anuchapreeda, and Yingmanee Tragoolpua. "Anti-herpes simplex virus activity of extracts from the culinary herbs *Ocimum sanctum* L., *Ocimum basilicum* L. and Ocimum americanum L." *African Journal of Biotechnology* 10, no. 5 (2011): 860–866.

Zhang, Ji-Wen, Sheng-Kun Li, and Wen-Jun Wu. "The main chemical composition and in vitro antifungal activity of the essential oils of *Ocimum basilicum* Linn. var. pilosum (Willd.) Benth." *Molecules* 14, no. 1 (2009): 273–278.

Zheljazkov, Valtcho D., Charles L. Cantrell, Babu Tekwani, and Shabana I. Khan. "Content, composition, and bioactivity of the essential oils of three basil genotypes as a function of harvesting." *Journal of Agricultural and Food Chemistry* 56, no. 2 (2007): 380–385.

Zoghbi, Maria das Graças B., Jorge Oliveira, Eloisa Helena A. Andrade, José Roberto Trigo, Roberta Carolina M. Fonseca, and Antonio Elielson S. Rocha. "Variation in volatiles of *Ocimum campechianum* Mill. and *Ocimum gratissimum* L. cultivated in the North of Brazil." *Journal of Essential Oil Bearing Plants* 10, no. 3 (2007): 229–240.

Zollo, PH Amvam, L. Biyiti, F. Tchoumbougnang, C. Menut, G. Lamaty, and P. H. Bouchet. "Aromatic plants of tropical Central Africa. Part XXXII. Chemical composition and antifungal activity of thirteen essential oils from aromatic plants of Cameroon." *Flavour and Fragrance Journal* 13, no. 2 (1998): 107–114.

Index

D

E

F

G

H

I

J

K

L

M

N

O

P

Q

R

S